WINDOWS ON THE LIFE SCIENCES

BY

RICKI LEWIS, PhD

THE UNIVERSITY AT ALBANY

CARENET MEDICAL GROUP

SCHENECTADY, NEW YORK

b

Blackwell
Science

Editorial Offices:
Commerce Place, 350 Main Street, Malden, Massachusetts 02148, USA
Osney Mead, Oxford OX2 0EL, England
25 John Street, London WC1N 2BL, England
23 Ainslie Place, Edinburgh EH3 6AJ, Scotland
54 University Street, Carlton, Victoria 3053, Australia
Other Editorial Offices:
Blackwell Wissenschafts-Verlag GmbH, Kurfürstendamm 57, 10707 Berlin, Germany
Blackwell Science KK, MG Kodenmacho Building, 7-10 Kodenmacho Nihombashi, Chuo-ku, Tokyo 104, Japan

Distributors:
USA

Blackwell Science, Inc.
Commerce Place
350 Main Street
Malden, Massachusetts 02148
(Telephone orders: 800-215-1000 or 781-388-8250; fax orders: 781-388-8270)

Canada

Login Brothers Book Company
324 Saulteaux Crescent
Winnipeg, Manitoba R3J 3T2
(Telephone orders: 204-224-4068)

Australia

Blackwell Science Pty, Ltd.
54 University Street
Carlton, Victoria 3053
(Telephone orders: 03-9347-0300;
fax orders: 03-9349-3016)

Outside North America and Australia

Blackwell Science, Ltd.
c/o Marston Book Services, Ltd.
P.O. Box 269
Abingdon
Oxon OX14 4YN
England
(Telephone orders: 44-01235-465500;
fax orders: 44-01235-465555)

Acquisitions: Nancy Hill-Whilton
Development: Jill Connor
Production: Erin Whitehead
Manufacturing: Lisa Flanagan
Director of Marketing: Lisa Larsen
Marketing Manager: Carla Daves
Cover Design: Meral Dabcovich
Typeset by Software Services
Printed and bound by Edwards Brothers/Ann Arbor

Printed in the United States of America
00 01 02 03 5 4 3 2 1

Library of Congress Cataloging-in-Publication Data
Lewis, Ricki.
Discovery : windows on the life sciences / by Ricki Lewis.
p. cm.
ISBN 0-632-04452-7
1. Molecular biology. 2. Biology, Experimental. 3. Biochemistry. I. Title.

QH506 .L4365 2001
572.8—dc21 00-034229

To Larry, Heather, Sarah, and Carly

TABLE OF CONTENTS

PREFACE

Science as it comes to the average citizen often lacks any of the excitement of discovering something in or about the natural world. The evening news discusses a complex disease in a scant few minutes, with time only to shower accolades on one or two researchers, talk about cures, and, of course, invoke the word *breakthrough.* A story of biomedical science in a prominent newspaper can send stocks soaring and patients running to their doctors demanding treatments that could, in reality, be decades away. When a new biotechnology transcends the scientific literature, presidents belatedly appoint bioethics committees, and elected officials make statements and call for restrictions in research funding when what they really need is a crash course in basic biology.

Lost in the media fuss, and overshadowed by pervasive sciencephobia, is the fact that science proceeds by steps so incremental that they come to form a continuum, with contributions occurring over many years, from many individuals. This collection of essays attempts to place some of the life science topics that dominate today's headlines, and a few that are still unnoticed, into perspective, tracing the roots of knowledge, how discovery happens, and how science spawns technology.

I have been writing science news and feature articles for a variety of publications, as well as biology textbooks, for over 20 years. Both outlets are satisfying, yet frustrating in the same way—they present tips of icebergs. Rarely does a magazine or newspaper article have the time or space to devote to the evolution of ideas that drives scientific research. And today's fact-packed biology textbooks must set aside the stories of how we have come to know what we know, considering them to be so much padding, unnecessary fluff. A textbook in today's market must stick to the facts. But how the facts got there is often as interesting as what they say about life.

Discovery has allowed me the freedom to select a few topics of current great interest in the life sciences and to probe their origins and evolution. Cloning research did not just suddenly begin with the birth of one sheep in 1996, nor did stem cells burst into consciousness two years later. These particular topics are media darlings. Others, such as prions and telomeres, make only occasional appearances to the public, typically in the guise of brain disease or everlasting life, respectively; instead, they primarily stimulate excitement within the circle of biologists.

Although each essay stands alone, their total organization follows a familiar conceptual flow from small to large: from the biochemicals of life to cellular matters, and then to development and the human organism. Sandwiching these essays are an introductory one that discusses different modes and circumstances of scientific discovery, and a final essay that probes, via the examples of gene therapy and genetically modified foods, difficulties that arise when science becomes technology.

Delving into the detailed stories of discovery reveals some intriguing personalities, those whose insight clearly drives progress, yet many whose contributions have been bypassed. My research has uncovered a few such "unsung heroes." Little-known Russian theoretical biologist Alexey Olovnikov hypothesized a link between shrinking chromosome tips and cellular aging before anyone else, influenced by an observation about subway trains. The Jackson Laboratory's Leroy Stevens laid the groundwork for stem cell biology from the 1950s through the 1970s by noticing, and pursuing, an intriguing strain of mice. Also in the 1950s, Icelandic pathologist Bjorn Sigurdsson gave a series of lectures at the University of London describing a mysterious disease in sheep that he called *rida*; these lectures were early descriptions of prion diseases, which later came to be associated with D. Carleton Gajdusek and Stanley Prusiner. And Prusiner, along with Carl Woese and his beautiful discovery and description of the third type of life, the archaea, exemplifies the rare breed of scientist whose once-renegade ideas, over time, come to define true paradigm shifts.

The style of the essays is a hybrid of journalism and textbookese, perhaps because I am a hybrid journalist/scientist. I've let the researchers themselves, or their colleagues, tell the stories wherever possible. For each essay one or two "gurus" emerged—sources who took a special interest in this project and nurtured it, reading drafts, making suggestions, and easing my contact with other sources.

For the first essay, which sets the tone for the diversity of ways in which discoveries happen, Neil Smalheiser introduced me to the story of John Cade and the convoluted investigation of lithium's effects on the brain. Stanley Miller happily and helpfully retold his tale of being portrayed in the press as creating life in a test tube for Chapter 2. Carl Woese spent hours detailing his surprise at finding the archaea by looking in places that no one had before, and his frustration at the time it has taken for his work to be fully accepted. Roulette William Smith told me of Sigurdsson's early prion work, and Thomas James explained his imaging experiments of prions, reading several drafts of that chapter. Leonard Hayflick, Alexey Olovnikov, Carol Greider, and Jerry Shay patiently explained the chronology and credits that are part of the telomere story. I found the founder of stem cell research, Leroy Stevens, simply by following the literature back in time, supplementing the paper trail with interviews with Stevens' coworkers and daughter—he suffered a stroke about 10 years ago, but nevertheless read the draft chapter and offered suggestions. Dennis Steindler served as the present-day stem cell guru, reviewing several drafts and supplying many fascinating papers as background. My on-call cloner was Steve Stice, always answering my many questions promptly and expertly. And for alerting me to the homocysteine story, I thank Paul Swartz for introducing me to the trials and tribulations of his college roommate Kilmer McCully and his efforts to show that cholesterol isn't the only risk factor for heart disease. Laurence Pilgeram alerted me to his rarely mentioned early work in that field, and James Finkelstein painstakingly straightened out the media-twisted tale of who discovered what and when in the homocysteine story.

I also thank Margaret Simpson, Ken Renner, Scott Mohr, Leos Kral, Christine Genovese, and Jack Fabian for reading selected chapters in their areas of expertise. A special thanks to Sandy Latourelle, who tirelessly read and commented on every essay, and I thank my husband, Larry, and three daughters for helping me through tough times during the writing. Finally, I thank Barry Palevitz for originating the idea for this book.

Ricki Lewis

ON DISCOVERY

Discovery in science happens in many ways. Introductory biology courses can give the impression that a scientist goes into the lab every morning as if it were a 9-to-5 job, armed with a "to do" list for the day—observation, hypothesis, experiment, results, and conclusions. But as anyone who has explored nature knows, science rarely, if ever, proceeds by a process as rigid and ordered as "the" scientific method. Hunches, intuition, gut feelings, and sheer luck have as much to do with discovery as careful planning of experiments. And despite the frequently evoked phrase "scientific proof," no scientific conclusion is really conclusive. New observations and experimental results enable scientists to continually reject and refine ideas, to offer the best explanations, and to ask further questions.

Yet exploring nature can be methodical too, such as cataloging the species that live in a rainforest's treetops, searching body fluids for the source of a newly recognized infectious disease, or sequencing DNA. Some experiments are carried out just to see if we can do something we've imagined, such as cloning a mammal (see Chapter 7). In still other types of scientific inquiry, indirect clues and the work of many others tell a researcher that a certain something must exist, and then clever experiments teamed with sharp analytical tools identify what the scientist strongly suspects is there. Elizabeth Blackburn and Carol Greider took this approach to identify telomeres at chromosome tips (see Chapter 5), Stanley Prusiner did so to isolate the infectious proteins he named prions (see Chapter 4), and Watson and Crick combined clues from others' experiments to reveal the DNA double helix that Gregor Mendel could hardly have imagined.

Some experiments try to unravel nature by building complex systems from components. Researchers have put together the genes of a virus, built chromosomes, and activated genes that turn a cell cancerous. But this bottom-up approach can have limits when time enters the equation, such as in attempting to recreate how life began. We can, theoretically, combine the very molecules that long ago assembled and interacted to form the first protocells, but we can never know if we have really replicated life's origin (see Chapter 2). Nor can we predict the future, no matter how well we understand the natural laws of the present.

Because science is both an art and a science, scientists come in two varieties too, with the best embracing both approaches. Mogens Schou, a psychiatrist at the Aarhus University Institute of Psychiatry in Risskov, Denmark, defines these two types of scientists as "systematic" and "artistic." Schou did much of the "systematic" type of work following the discovery of lithium as a treatment for manic depression in 1949, discussed later in this chapter. The systematic scientist works in a step-by-step fashion from point A to point B to point C to point D. In contrast, the artistic scientist has a "feel" for research, following intuition and logic and just letting the ideas flow and roam, even when the connections seem tenuous, the experiments risky. The artistic scientist takes the obvious A to B to C to D trajectory just to be safe, but is not afraid to go beyond to points E and F or to find alternate routes (Figure 1.1).

No scientific discovery is truly serendipitous, for to suggest that a revelation just falls into a scientist's lap is rather insulting. Alexander Fleming had to have known something about bacteria and fungi to realize that the clear areas he saw on culture plates signaled the presence of an antibiotic compound. Horace Walpole coined the term "serendipity" in 1754 after reading a fairy tale called "The Three Princes of Serendip," a location that is an ancient term for Ceylon, today known as Sri Lanka. The trio of princes made a series of purely fortuitous discoveries, but Walpole extended his definition of serendipity to mean "unexpected discoveries by accidents and sagacity." Many years later, Louis Pasteur made a similar, and more widely quoted, observation: "Chance favors only the prepared mind."

The accounts of scientific discovery in this chapter and throughout the book run the gamut from an observation seemingly coming from thin air and leading to sudden inspiration to a direct experiment aimed at answering one question that instead raises or answers another, and from following seemingly outrageous ideas to undertaking large-scale quests, such as combinatorial chemistry and sequencing genomes, that are nearly guaranteed to produce results.

The Systematic Scientist

A → B → C → D

The Artistic Scientist

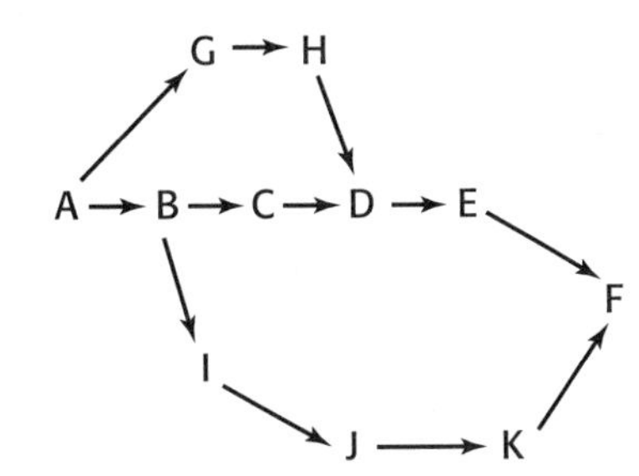

FIGURE 1.1

A systematic scientist proceeds through a linear series of steps, whereas an artistic scientist takes detours or seeks alternate pathways. Of course, many scientists use both approaches, depending upon the questions he or she seeks to answer.

SERENDIPITY

In 1965 Jim Schlatter, a young researcher at a pharmaceutical company, was trying to synthesize a drug to treat ulcers. One day, he was attempting to make an intermediate compound, a chain of four amino acids. While working, his hand accidentally brushed against the rim of a flask, picking up a tiny amount of fine white powder on a fingertip. When Schlatter licked the finger briefly a few minutes later to pick up a piece of paper, he was startled by the incredibly sweet taste. Where had it come from?

Without thinking about the potential danger of eating an unidentified white powder, Schlatter retraced his steps to the flask, and on close inspection spotted the chemical. He collected a sample and performed several standard analytical tests to find out what it was. He also tasted it, this time intentionally, to confirm that it was the same sweet substance he had sampled earlier. The sweet substance turned out to be a surprisingly simple compound, very similar to the amino acids aspartic acid and phenylalanine. Because protein is a nutrient, Schlatter realized that what he had found was in all likelihood safe to eat, at least in the short term—the fact that he was still alive and well indicated that this was so. The constituent amino acids alone produce either no taste or a bitter taste; their sweetness when linked and modified was completely unanticipated.

Nine years and many experiments later, the U.S. Food and Drug Administration (FDA) approved the new sweetener, called aspartame. It is 200 times as sweet as sucrose (table sugar), and the body rapidly converts it to the two natural amino acids. Aspartame is low in calories and safe for use by people with diabetes. Although the Internet is active with anecdotal accounts of excess aspartame causing headaches in some people, clinical trials have not borne this out, and the sweetener remains in a wide variety of foods and is safe for most individuals. (However, it is dangerous for people with the inborn error of metabolism PKU, and it is possible that some of the reports of adverse effects come from people who are carriers of PKU, information which isn't typically known.)

Jim Schlatter discovered aspartame while searching for a treatment for ulcers. A different study of ulcers illustrates another aspect of scientific discovery, namely, that long-held ideas, no matter how logical, can be wrong. For decades, painful gastric ulcers were blamed on the stress of a fast-paced lifestyle that triggered excess stomach acid secretion. As long ago as 1910, a noted German physician claimed "no acid, no ulcer." In the 1970s, when acid-lowering drugs were found to be effective, if only temporarily, in alleviating ulcer symptoms, the link was strengthened. For many people, gastric ulcers thus became a lifelong condition, treated with a bland diet, stress reduction, acid-lowering drugs, and sometimes surgery. But stress does not directly cause many ulcers—a bacterial infection does.

In the early 1980s, J. Robin Warren, a medical researcher at Royal Perth Hospital in western Australia, was investigating bacteria that lived in the stomach linings of people suffering from gastritis. Because this inflammatory condition often precedes the appearance of ulcers, Warren wondered if the bacteria played a causative role. Or did they just seek stomachs already riddled with ulcers? He was intrigued at the ability of the bacteria to not only survive in, but thrive in, the highly acidic conditions of the human stomach.

While Warren was figuring out how to test his hypothesis that the bacteria caused the ulcers, his impatient young assistant, medical resident Barry Marshall, came up with an experiment. On a July day in 1984, he brewed what he called "swamp water"—a solution of a billion or more bacteria—and drank it. Marshall had a healthy stomach, but after swallowing the swamp water he soon developed gastritis. He was lucky, though. His symptoms cleared on their own and did not return. Another assistant who self-experimented suffered for months, finding relief only after a course of antibiotic drugs.

Other researchers repeated Marshall's experiment on laboratory animals, giving them bacteria-laden drinks, and eventually confirmed that *Helicobacter pylori* causes gastritis as well as ulcers. Still, it wasn't until 1994 that the FDA officially alerted doctors to add antibiotics to the regimen to treat gastric ulcers. Today, entire medical journals and conferences are devoted to *H. pylori,* and at a recent annual meeting of the American Society for Microbiology, a drug company that manufactures antibiotics handed out stuffed "animals" of the bacteria.

FOLLOWING A HUNCH: THE ROAD TO CISPLATIN

The April 15, 1999, issue of the *New England Journal of Medicine* brought great satisfaction to Barnett Rosenberg, a retired Michigan State University biophysicist. Not many scientists can watch basic research blossom into a cancer treatment that has extended or saved thousands of lives. Rosenberg's keen observation in 1961 led, 18 years later, to FDA approval of cisplatin. The drug's conception, gestation, and birth illustrate the value in following even the strangest of hunches.

Cisplatin revolutionized the treatment of testicular cancer and is effective against some ovarian and colon cancers too. Three reports in the 1999 journal added cervical cancer to the list. But the road to cisplatin's discovery and development was hardly smooth, and the tale offers insight into the challenges of basic research and the importance of intuition and persistence.

The story began in 1961. Michigan State University in East Lansing had recruited Rosenberg from New York University to start a biophysics department, although he claims he had "no competence in biology—few physicists have." Looking for biologically oriented research questions, he decided to follow up on an observation. He had long wondered about the resemblance of iron filings clinging to a bar magnet to the appearance of condensed chromosomes in a cell caught in the throes of division. "In my earlier reading I had been fascinated by the photomicrographs of the mitotic figures in cells in the process of division. They called to a physicist's mind nothing so much as the shape of an electric or magnetic dipole field, the kind one sees with iron filings over a bar magnet," Rosenberg wrote in *Interdisciplinary Science Reviews* in 1978. How might a magnet affect cells? he wondered.

Rosenberg decided to use an electromagnetic setup with platinum electrodes, a metal known to be inert. But before he added mammalian cells whose chromosomes he wished to watch, he used smaller and easier-to-grow bacterial cells to set the parameters of the experiment. He turned on the electromagnet once the bacterial culture had reached a steady state in which new cells formed at the same rate at which old ones died. But soon after turning on the electricity, the microbial population plummeted. When he turned the electromagnet off, bacterial reproduction resumed. That in itself was interesting, but when Rosenberg examined the bacteria, he was startled: After exposure to the electromagnet, some had extended to 300 times their normal length! Apparently the cells had grown, but not divided.

The implications were clear—Rosenberg had found something that blocks cell division, and such a something might counter the out-of-control cell cycle that is cancer. But it wasn't the electrical field that stopped the bacterial cells from dividing, an obvious explanation.

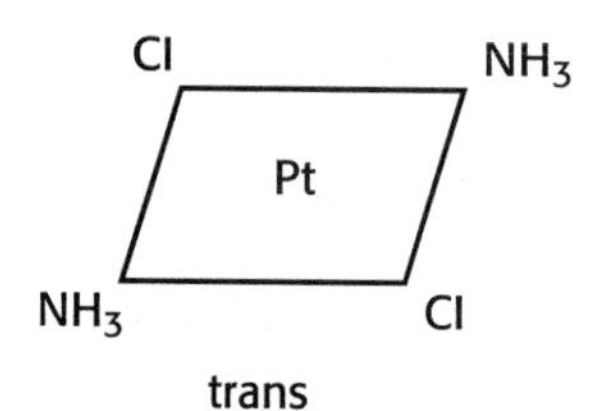

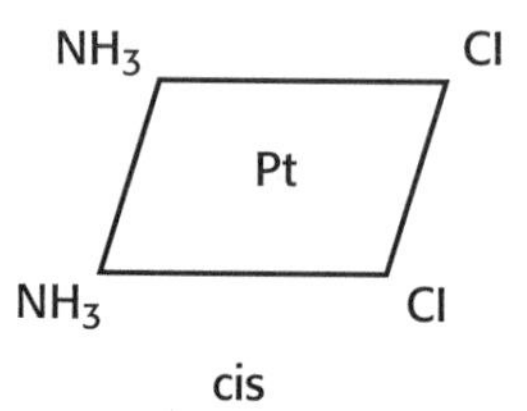

FIGURE 1.2

Isomers of cisplatin. The *trans* form has the chlorines kitty-corner as well as the ammonias in this configuration. The *cis* form has the chlorines, and the ammonias, each on the same side of the molecule.

After conferring with chemists, Rosenberg and his colleagues realized that a chemical reaction had occurred between the ammonium chloride used as an electrolyte in the fluid that bathed the electrodes, and the platinum. The product of the reaction was a compound called dichlorodiamineplatinum.

In science one experiment typically follows another, sometimes even without conscious planning, as each triggers new questions. Research in the Rosenberg lab now veered from biophysics to chemistry. But synthesizing the platinum compound and adding it to bacterial cultures did not have the anticipated effect of generating giants. Sharp observational skills helped again. The researchers noted that dichlorodiamineplatinum left to sit on a shelf for a few weeks did have the desired effect of blocking bacterial cell division. How was the compound on the shelf different from fresh material? The answer was exposure to light. Dichlorodiamineplatinum had undergone a photochemical reaction.

The confusion continued, however. Although the platinum compound freshly synthesized in the lab had no effect on bacteria, it indeed halted cell division when collected from the electrodes. It turned out that dichlorodiamineplatinum exists in two forms, or *isomers* (Figure 1.2). The *trans* form has the two chlorines and two ammonias kitty-corner in the molecule, and the *cis* form has them on the same side.

Because the *trans* form is more stable, it predominates in a synthesis reaction, but it is the elusive *cis* form that disrupts bacterial cell division—and only when present in large amounts. In addition, photochemical reactions tend to favor formation of the isomer that is less likely to form in a synthesis reaction, and so exposure to ultraviolet wavelengths on the shelf converted *trans* molecules to *cis.* And so both the compound that formed on the electrodes and the one that formed after exposure to light consisted of predominantly the *cis* isomer. It was the excess of *cis*-dichlorodiamineplatinum—known as *cisplatin* in pharmaceutical parlance—that halted bacterial cell division. When Rosenberg searched the scientific literature, he discovered that chemists had synthesized the compound in 1848 and called it "Peyrone's chloride," and again in 1890. Plus, chemists had described antibacterial properties of platinum-group metals in 1953.

Now the artistic scientist put on his systematic scientist's hat. Rosenberg (and others) screened the effects of various platinum-containing compounds on bacteria, and looked at the responses of various types of bacteria to cisplatin. They discovered that only a high concentration of the *cis* form of cisplatin stopped cell division. Moreover, treated bacterial cells recover. This was unlike their response to nitrogen mustard, a chemical that paralyzes DNA synthesis. Cisplatin, therefore, does not work by halting DNA replication.

In 1968 the Michigan State researchers performed crucial LD_{50} tests, a standard measurement in toxicology that determines the dosage necessary to kill half of a group of test animals. They performed tumor-shrinkage tests in mice. In the protocol, repeated many times, mice received transplants of a standard sarcoma (a solid tissue tumor) on day zero, cisplatin or placebo on day one, and then were sacrificed on day eight. Results were spectacular. Recalls Robert Switzer III, president of NeuroScience Associates in Knoxville, Tennessee, "I was in the lab during those heady times when the mouse-sarcoma results were unfolding. Little did I know of the full significance it all would bring—nor how it would endure."

Rosenberg felt compelled to notify the National Cancer Institute (NCI) immediately of his findings, so that people with cancer might be helped as soon as possible. The powers-that-be at the agency invited him to present his findings in a talk, which he recalls was met with "perceptible but understandable coolness." Potential cancer cures were, after all, a dime a dozen. But he wisely left several samples of cisplatin for government researchers to incorporate into their drug screening program. Sure enough, when the NCI researchers tested cisplatin on mice with leukemia, not only did the drug work on this second type of cancer, but tweaking the protocol to give the drug on day eight achieved a cure rate close to 100%. To put this into perspective, a new cancer drug that helps 15% of patients is considered reason for celebration.

With this success, Rosenberg's National Cancer Institute grant was funded, and in 1969, the journal *Nature* published the first report on cisplatin's anticancer activity. "An American journal of almost equal distinction turned it down," recalls Rosenberg, because a reviewer claimed that cisplatin was hardly noteworthy among the many new cancer drugs being described. Perhaps the lack of attention in some circles was a sign of the times, for cisplatin's discovery predated President Richard M. Nixon's official declaration of "war" against cancer in 1973. (Of course, cancer research didn't suddenly bloom into existence then.) But when professor Sir Alexander Haddow, then head of the Chester Beatty Institute in London, showed that cisplatin was active against melanoma in mice, more people took notice. If the drug tamed three cancers—sarcoma, leukemia, and now melanoma—it was probably attacking a fundamental characteristic of the deranged cells.

Research intensified. Tests on dogs and monkeys were used to determine doses and toxicities of the drug. Molecular studies revealed that the *cis* compound crosslinks DNA that is rich in guanine and cytosine (two of its four types of building blocks), both within and between strands. This action essentially gums up the chromosomes of dividing cells, including bacteria, cancer cells, and certain normal cells that divide frequently.

In 1972 clinical trials began on cancer patients who had exhausted all standard treatments, providing information that would be used to establish dosages and side effects. Surprisingly, for there isn't much hope in such investigations, 20% of the participants had partial or complete remissions for varying times! But kidney toxicity was emerging as a hurdle, causing one reviewer to comment, "This is too interesting a drug to drop, but too dangerous for extensive use."

Work continued. Phase II clinical trials evaluated patients at earlier stages of disease, with each patient receiving cisplatin only if another drug failed. Meanwhile, animal studies revealed synergies of cisplatin with other drugs and radiation, effects echoed in the 1999 reports of success against cervical cancer. Other studies revealed that cisplatin inhibits repair of radiation damage by a circuitous route: It shrinks the tumor, which allows more oxygen to reach the cells. As the cells resume the division cycle, they become more sensitive to the radiation.

A key pilot study conducted at the Roswell Park Memorial (now Cancer) Institute in Buffalo, New York, was critical to the drug's approval in 1979. Thirteen of 15 men with testicular cancer responded to some degree for long time periods, and 7 were completely cured. Bristol Laboratories in Syracuse, New York, began producing cisplatin, which was added to protocols to treat cancers of the testes, ovaries, and bladder. Production of a related drug, carboplatin, soon followed.

Over the years, other drugs have been added to protocols that counter cisplatin's side effects, and more than a dozen clinical trials are ongoing to explore the drug's effects on other types of cancer. But much of the success of cisplatin stems from Barnett Rosenberg's curiosity and perseverance. Recalls Switzer, "Barney's mind was about as fertile a ground for ideas as anyone I've ever met. His enthusiasm and energy are infectious. Above all, he believes in his ideas and will ride them 'til they succeed or die. He could have 'bailed' any time in the face of the skepticism he faced."

SEEING CONNECTIONS: FROM LETHARGIC GUINEA PIGS TO LITHIUM

Sometimes a scientific discovery is unfairly attributed to serendipity. This was the case for the discovery that lithium can relieve and prevent the mood swings of bipolar affective disorder, more commonly known as manic depression. Lithium treatment has vastly improved the lives of thousands of people, but its discovery stemmed not from a lucky accident but from a researcher's astute observations and his ability to see subtle connections.

In 1949, Australian psychiatrist John Cade reported the results of a small-scale clinical trial of lithium carbonate in treating manic depression. A man of eclectic interests (such as studying the color patterns of magpies and comparing droppings of various Australian animals), Cade wrote papers in the decidedly nonscientific style of the first person that nevertheless eloquently captured his logical thinking and the excitement of his findings. Yet at the same time, he continually downplayed his success, describing himself as "an unknown psychiatrist, working alone in a small chronic hospital with no research training, primitive techniques and negligible equipment . . . ." It was not an exaggeration.

The bizarre symptoms of mania fascinated Cade—the restlessness, the sloppiness, the hyperactivity and mischievous nature of the sufferers. So did the opposite extreme of depression. Might the cause of mania and depression be similar to that of hyperthyroidism and hypothyroidism—too much and too little, respectively, of a single substance, he wondered? Cade came up with this idea as a prisoner of war of the Japanese during the second World War, when he had 3 years alone to try to keep his mind occupied. The idea led directly to a hypothesis: If mania and depression represented the excess and deficit of a molecule, then people with mania should excrete the excess. The hypothesis would turn out to be incorrect; that is, mania does not produce a measurable substance in the urine. But the investigation led to a totally unexpected and ultimately very useful discovery.

The experiment that sprang to Cade's mind was to inject laboratory animals with urine from people with mania. He was able to do so, on the sly, in a back room at Bundoora Repatriation Hospital. Controls consisted of urine from healthy people and from depressed people. He found that urine from people with mania injected into guinea pigs induced a characteristic syndrome at much lower doses than did the control urine—a sign of greater toxicity. The guinea pigs treated with "manic urine" seemed fine for 12 to 18 minutes, then trembled and staggered about. Within minutes, they became paralyzed, although they remained conscious. Then the animals began to twitch and convulse, lost consciousness, and died. It was an all-or-none response, but every once in awhile, a guinea pig survived.

Determining which component of urine harmed the guinea pigs was technically challenging. The culprit appeared to be urea, but a related waste product, uric acid, needed to be present for urea to be toxic. To isolate the effects of uric acid, Cade needed to dissolve it, and as anyone with urinary stones knows, uric acid is mighty insoluble in water. So Cade synthesized the lithium salt of uric acid, which increased the solubility to a workable level. To his great surprise, the lithium salt made the urea considerably less lethal. He had attempted to demonstrate an effect from urea, and had instead found it in uric acid.

To see if the effect arose from the lithium, he next tested urine that contained lithium carbonate, and saw the same result: Urine with an 8% urea solution killed 50% of guinea pigs, but the same solution plus 0.5% lithium carbonate killed none. "This argues for a strong protective function for the lithium ion against the convulsant mode of death produced by toxic doses of urea," Cade concluded.

The next experiment was the crucial one. What would happen if he injected lithium ion into guinea pigs? Only someone who has had these appealing animals as pets (or laboratory specimens) can appreciate what he saw. After a 2-hour latency period, the animals became lethargic for 2 hours. Sitting like hairy lumps is not unusual for guinea pigs, but the animals' apparent apathy certainly was. Guinea pigs are by nature skittish, backing away from a friendly hand, startling easily, and panicking at being placed on their backs. "Those who have experimented with guinea pigs know to what degree a ready startle reaction is part of their makeup. It was thus even more startling to the experimenter to find that after the injection of a solution of lithium carbonate they could be turned on their backs and that, instead of their usual frantic righting reflex behavior, they merely lay there and gazed placidly back at him," said Cade at a symposium on biological psychiatry in April 1970. By the end of the second 2 hours, the guinea pigs were back to their typical jumpy selves.

Cade made an intellectual leap: Lithium calmed guinea pigs given urine from people with mania. If the urine created an excess of some behavior-altering substance, did the lithium counter it?

The next experiment would hardly have made it past an institutional review board today because the human clinical trials were conducted much too soon for safety's sake. For the first test, Cade joined Jim Schlatter, Barry Marshall, and a long list of other scientists who have experimented on themselves (see Box 1.1). In separate experiments, he swallowed lithium citrate and lithium carbonate, discovering that these salts did no harm, but that the carbonate salt made him nauseous. The next step was to give lithium citrate to patients.

Over the course of a year, Cade accumulated data on 19 individuals: 10 with mania, 6 with schizophrenia, and 3 with chronic psychotic depression. He wrote and spoke often about the dramatic turnaround that he witnessed in his first patient, a 51-year-old man suffering from mania for 5 years at the time of treatment. The man had been banished to a back ward because he was always getting into trouble and was dirty and abusive to hospital staff. On March 29, 1948, the man began taking lithium. By the fourth day, his therapist thought he sensed a slight change in the patient's demeanor, but was hesitant to report it. By the fifth day, though, others noted improvements—the man's characteristic disinhibition had improved, he was neater, could concentrate, and appeared calmer.

BOX 1.1

EXPERIMENTING ON ONESELF: DISCOVERING LSD

Lysergic acid diethylamide, better known as LSD, is a hallucinogenic drug that rose to fame in a synthetic form in the 1960s, but is naturally produced from a fungus that infects certain grain plants, such as rye. Bread made from infected grain can cause the frightening symptoms of ergotism, including uncontrollable movements and terrifying visions, such as being possessed by the devil or chased by snakes. Ergotism may have been the impetus for burning women at the stake in colonial New England because people thought that they were witches. Known also as St. Anthonys' fire, which refers to a burning sensation in the hands, ergotism killed thousands in Europe in the Middle Ages.

LSD was first synthesized in the laboratory in 1938. Albert Hofmann, a chemist at Sandoz Laboratories in Basel, Switzerland, was developing drugs to treat migraine headaches, and he was experimenting with alkaloids from ergot-producing fungi. Meticulous about recording his observations, Hofmann's notebook entry from Friday, April 16, was quite telling. Hofmann left for home about two hours early that day because he was dizzy. He felt "a not unpleasant delirium which was marked by an extreme degree of fantasy . . . fantastic visions of extraordinary vividness accompanied by a kaleidoscopic play of intense coloration continuously swirled around me." John Lennon would echo that description some 60 years later in the song "Lucy in the Sky with Diamonds," whose initials are intentional and whose lyrics mention "the girl with kaleidoscope eyes." When Hofmann returned to normalcy, he reviewed the day's events, recalling that he'd made some crystals of lysergic acid diethylamide. Could he have accidentally eaten or inhaled some of the compound?

To find the answer, Hofmann did the logical, if risky, thing—he purposely ate some and awaited the effects. He didn't have to wait very long. After 40 minutes, he wrote, he experienced "mild dizziness, restlessness, inability to concentrate, visual disturbance, and uncontrollable laughter." Soon after he could no longer write. With assistance, he managed to ride his bicycle home, where symptoms rapidly intensified. He became unable to talk and had the impression that he was not moving at all, when in actuality he was racing about. He compared his perceptions to looking in an amusement park mirror that greatly distorts images. Faces appeared as terrifying masks. He had a metallic taste in his mouth and a dry throat, and felt as if he couldn't breathe and was looking at himself from afar. Six hours after ingesting the LSD, the hallucinations were still strong; he wrote, "[W]ith closed eyes, multihued metamorphizing fantastic images overwhelmed me," and once again invoked the idea of seeing the world through a kaleidoscope.

Hofmann concluded that certain natural substances, in exceedingly small doses, can induce a temporary state like mental illness, schizophrenia in particular. From the time of Hofmann's discovery until the mid-1960s, Sandoz laboratories explored ways to alter or deliver LSD to treat certain mental illnesses. But in 1966, when people began abusing LSD, Sandoz gave all its supplies to the National Institute of Mental Health for research purposes and ceased manufacturing the drug. Research has only recently resumed; abuse remains strong. LSD may find legitimate therapeutic value in inducing labor, lowering blood pressure, and relieving the pain of migraine headaches—where Hofmann's work began.

By the end of 3 weeks, he was moved to a convalescent ward. On July 9, he was discharged, and was able to hold a job—for awhile. Six months later he returned to the hospital much as he was before, his relatives stating that he'd stopped taking the lithium 6 weeks earlier. The man had unwittingly contributed valuable clinical information: The treatment must be continuous, or at least not suddenly stopped, to prevent symptoms. Cade and others expanded the trials, and found that lithium treatment had no effect on people with depression, but it calmed those with schizophrenia sufficiently so that they did not need sedation.

Cade published "Lithium Salts in the Treatment of Psychotic Excitement" in the September 3, 1949, issue of the *Medical Journal of Australia.* Not many other physicians noticed. Part of the reason for the lack of fanfare was lithium's sordid reputation. In 1859 English physician and authority on inborn errors of metabolism Archibald Garrod had found that lithium could be used to treat the painful joint inflammation of gout. This led to its use to treat a variety of illnesses, but most of the applications were completely unfounded. By 1907, physicians were noting cases of lithium toxicity, but their observations didn't attract much notice until the 1940s, when its use to treat the edema of congestive heart failure led to several deaths, which were reported prominently in the *Journal of the American Medical Association.* It was against this historical backdrop, plus competition from other psychotropic drugs, that Cade's 1949 paper hailing lithium in treating mania appeared.

Interest in lithium did not expand beyond a few dozen publications a year until the 1960s, thanks largely to the work of Mogens Schou and other systematic scientists who rigorously tested the treatment. As for Cade, he quickly moved on to studying the effects of other ions on the human body, claiming, with typical modesty, that "I am not a scientist. I am only an old prospector who happened to pick up a nugget." But, like other artistic scientists, he did far more than stumble upon an answer to a question. His discovery came from making connections. But the precise mechanism of how lithium counterbalances whatever is abnormal in manic depression remains unknown.

SYSTEMATIC SEARCHES

Nature provides an incredible diversity of living and nonliving objects to explore. Some forms of scientific discovery entail searching through nature's bounty to find interesting or useful things. But this can be done in different ways.

NATURAL PRODUCTS CHEMISTRY

It is an oft-quoted statistic that the active ingredients of one-quarter of all prescription drugs come from plants. Visit a pharmacy or supermarket these days, and everything from drinks to cough drops to vitamin preparations also contain such "herbals" as ginkgo, garlic, and echinacea, promising effects that range from improved mental alertness to infection protection to a better sex life. The active ingredients of many other drugs are analogs created by chemists following cues from nature's bounty.

Today's chemists specializing in natural products consult a treasury of clues from history and anthropology, consulting the pharmocopeias of native peoples who discovered, by trial and error, which plant extracts had which effects. Consider the bark of the Pacific yew tree *Taxus brevifolia.* A screen of thousands of natural products for medicinal qualities conducted by the National Cancer Institute identified an anticancer compound from the Pacific yew in 1960. It took until 1992 for it to be approved to treat ovarian and breast cancer as the drug Taxol (paclitaxel). But native peoples had used extracts from the tree for other purposes for centuries. The Potawatomi eased the pain of syphilis sores with crushed leaves, and the Iroquois, Menominee, and Chippewa used steam from the leaves to relieve arthritis.

The discovery and development of Taxol illustrates a common sequence of events from nature to laboratory to drugstore. The bark of one century-old, 40-foot-tall Pacific yew yields only 300 milligrams of the drug. During the 1980s, concerned environmentalists defending the trees publicized equations such as three trees dying to save a human life. But soon biologists discovered Taxol in more common trees and in a bark fungus, and eventually, chemists figured out how to synthesize it (not an easy task—nature does not do things simply).

But the future of natural products chemistry is uncertain, struggling against a backdrop of disappearing species. Researchers are racing to discover nature's medicine chest in the face of the destruction of the rainforests that house much of life's diversity. Peter Raven, director of the Missouri Botanical Garden and perhaps the most outspoken scientist in the United States concerning the coming biodiversity crisis, explained the situation:

> Plants provide, directly or indirectly, food, medicine, a large proportion of the chemical substances that we use, shelter and clothing. Yet we are destroying them at a frightening rate, so that as many as 100,000 of the estimated total 300,000 species may be gone or on the way to extinction by the middle of the next century. If we do not take action now, by the end of the century we may have destroyed two-thirds of the plant species we currently use and enjoy. Half of all species are in the tropical rain forests, and they will diminish to 5 percent of their current area over the next 50 years. Three-quarters of their species will be driven to extinction, or be on the brink. As many as one-sixth of them are currently unknown—and so we are losing what we have not recognized.

The living world offers an enormous number of compounds to scrutinize. Finding them has been dubbed "bioprospecting." Says Jim Miller, associate curator at the Missouri Botanical Garden, "There is tremendous excitement at this time in bioprospecting. Yes, there's a sense of urgency as resources are threatened, but potential for a huge number of discoveries. We are now surveying plants that humans have a history of eating, and looking for activities that could affect human health."

Such searches, though, are time-consuming and full of dead ends, as thousands of compounds are tested for activities first on cells in culture, then in experimental animals, and, ever so rarely, in clinical trials with humans (Table 1.1). "Bioprospecting is not a field that is punctuated by a series of 'eureka' moments. It is a slow, prodding process. Nevertheless, we are discovering natural products at a quick rate. Most will not go directly from a plant in the forest to a capsule, but will serve as model compounds that give pharmaceutical chemists new ideas," explains Miller.

To get an idea of the long journey from forest plant to pharmaceutical, consider the ongoing investigation of molecules called biflavones found in the fruit of a West African tree, *Garcinia kola*. West Africans have eaten the "nuts" of the bitter kola tree for centuries, and often include them in welcome baskets or as gifts because of their purported healing qualities as a "general antidote." Found in any fruit market, "it is used to treat nonspecific infection, where the healer does not know exactly what is being used," says Maurice Ewu, who is trained in pharmacognosy (the study of medicines derived from natural products) and is descended from a long line of Nigerian healers. Table 1.2 lists some of the uses of the kola tree.

TABLE 1.1

Steps in Traditional Drug Discovery

1. Find novel organism with interesting biochemical.
2. Extract biochemical from organism.
3. Screen extract for specific activities (antitumor, anti-inflammatory, antibacterial).
4. Isolate active compound(s).
5. Purify active compound(s).
6. Test active compound(s) for specific activities.
7. Check to see if active compound(s) is already described in literature or patented.
8. Develop drug formulation.
9. Conduct animal tests.
10. Conduct human clinical trials (safety and efficacy, dosage, postmarketing surveillance).

TABLE 1.2

West African Uses of *Garcinia kola*

Part of Plant	Use
Dried peel	Cuts and sore throats
Root soaked in gin	Tooth decay, cough, gonorrhea
Tea (steamed leaves)	Stomachache, hepatitis, respiratory tract inflammation, diarrhea, aphrodisiac
Seed	Liver cirrhosis, fertilization stimulant
Dried seed	Asthma
Fresh fruit	Bronchodilator

Ewu founded a nonprofit organization, the Bioresources Development and Conservation Programme, to test extracts of *Garcinia kola* in 1989. In vitro tests identified 46 promising compounds, which researchers then narrowed further to the flavonoids, compounds related to those found in citrus rinds. The tests identified quite a few "anti" activities against disease-causing organisms, including the bacteria *Streptococcus mutans* (a cause of tooth decay), *Bacillus subtilis, Shigella flexneri, Escherichia coli, Pseudomonas aeruginosa, Staphylococcus aureus,* and beta-hemolytic *Streptococcus.* Also on the list were the fungi *Aspergillus* and *Candida albicans,* as well as the single-celled organism that causes leishmaniasis (Chagas' disease). Extracts of *Garcinia* seemed to cure everything from pattern baldness to hangnails, and standard laboratory residents such as mice and rats could eat it for weeks unscathed—not surprising, since Africans regularly consume the plant. And because the compounds did not seem to disturb liver cells growing in culture, Ewu concluded that they are unlikely to have the liver toxicity that is the kiss of death to many would-be drugs.

Despite these promising results, it will be a long road toward drug development. The probability that these compounds will actually make it as far as the drugstore is about the same as that for any natural product undergoing testing—about 1 in 1000 to 10,000. More likely is that learning the structure of the natural product and how it forms will give chemists clues that they can use to improve on nature—and perhaps discover something about chemical reactions in the process.

In the meantime, *Garcinia kola* extract is available as a food supplement by a company that is affiliated with the nonprofit organization doing the research. It is a peculiarity of U.S. law that drug approval requires years of rigorous testing, yet impure extracts of the organisms that ultimately yield drugs may be marketed as "food supplements." This means that people can buy bathtub loads without any guidance, as long as the advertising does not make any claims to cure a specific illness.

SYNTHETIC VERSUS COMBINATORIAL CHEMISTRY

While some natural products enjoy commercial success as food supplements, others serve as models for chemists to create new drugs. But the traditional field of chemical synthesis is giving way to large-scale efforts to scan many related compounds for drug activity (or other applications), a field called *combinatorial chemistry.* Painstaking chemical synthesis of a single target compound and combinatorial chemistry are two very different approaches to the same goal. Consider an example of a one-compound-at-a-time effort.

In 1995, researchers at a large pharmaceutical company discovered anticancer activity in a natural product isolated from a soil bacterium. They named the compound epothilone A, and before you could say "FDA," a race was on to synthesize it, as opposed to extracting and purifying it from the natural source. By the end of 1996, two research groups had reached the finish line, devising ways to combine chemical reactions in a series of steps that culminates with the desired product. The scientific journals made much of the accomplishment as well as the race, fueling the idea that scientific discovery is all about competition.

Building a complex molecule using chemical reactions is called *total synthesis.* It is perhaps the most creative of the branches of chemistry, requiring logic, experience, and trial-and-error experimentation to copy or improve on nature. Synthetic chemists have made some incredibly large and intricate biomolecules, including vitamins, antibiotics, and analgesics. Consider palytoxin, a neurotoxin from a soft coral. It is so large that it contains 100 places where the atoms can assume either of two different orientations, one the mirror image of the other. To synthesize the molecule, chemists had to get the orientations correct at each of the 100 spots or the compound would not have the desired effect on the nervous system.

Only about a dozen laboratories worldwide currently work on *total synthesis.* Declining interest in the field has several causes. One is historical. At the time of the origin of synthetic chemistry, early in the twentieth century, one goal was for the reactions along the pathway to reveal, indirectly, the three-dimensional structure of the final product, a little like seeing a picture emerge as parts of a puzzle are assembled. After 1950, introduction of analytical tools such as x-ray crystallography and then nuclear magnetic resonance imaging provided easier ways to deduce chemical structures.

Another reason why total chemical synthesis is becoming an unusual form of discovery is the pressure for research to be applied rather than basic. Researchers tend to use known chemical reactions to travel biosynthetic pathways, rather than exploring new ways to get to the destination that might take longer. As a result, basic discoveries of how chemical bonds form and break are on the wane.

Combinatorial chemistry is perhaps the most powerful deterrent to the more creative but narrower strategy of total synthesis. This relatively new technology systematically generates thousands or even millions of compounds at once, each distinguished from the others by minute, subtle differences, all variations of the same molecular theme. Then the compounds are tested for characteristics that might make them drug candidates. Parts of the "combi chem" process, such as screening for certain activities, are so repetitive that robots perform them. Large pharmaceutical companies have embraced combinatorial chemistry in a big way, cutting back on programs that synthesize analogs to natural products one by one. Such analogs, whether generated by a chemist's idea of how to tweak a molecule or by the mass production that is combi chem, seek higher activity and fewer side effects than current drugs.

Combinatorial chemistry is a needle-in-a-haystack approach in which the scientist controls the composition of the haystack. But can it replace the millions of years of natural selection that have molded the chemicals of life in unpredictable, and sometimes unfathomable, ways? Combinatorial and synthetic chemistry are each routes to discovery, but they are as different in approach as Mogens Schou's systematic scientist, who plods methodically from point A to B to C to D, is from the artistic scientist who gets to D some other way. Yet there is room for synergy. Combinatorial chemistry can take cues from nature, beginning with natural products to generate its variations on a theme.

SEQUENCING GENOMES

Similar trends sometimes appear in different branches of science. In this sense, discovering natural products is to combinatorial chemistry as traditional genetics is to the new field of genomics. Natural products chemistry begins with an intriguing compound isolated from an organism. Likewise, traditional genetics begins with an interesting inherited trait in a particular type of organism. Just as chemists determine the natural product's structure and how it forms, so geneticists isolate and describe the gene that causes the interesting trait. Combinatorial chemistry generates and screens thousands of compounds at a time. The new field of genomics lays bare an organism's thousands of genes—its genome—and enables researchers to screen for particular activities and interactions among them, in different tissues and at different times in development, and to compare genomes in search of evolutionary clues.

The parallels between combinatorial chemistry and genomics apply in scale, but also in how some scientists perceive the creativity required for the old technique versus the new. Some argue that the new fields replace the one-gene-or-compound-at-a-time creative hunt with rapid-fire efficiency. Others claim that today's robotic crunching of data will supply a vast amount of raw material for scientists to mine in the new century and beyond.

A genome is the complete set of genetic instructions for an organism. The first to have their genomes sequenced were, logically, the smallest microorganisms. The bacterium *Hemophilus influenzae* led the way, making its debut in the pages of *Science* magazine in 1995 (Figure 1.3). J. Craig Venter, the founder and president of Celera Genomics Corporation in Rockville, Maryland, led the effort, and went on to lead many other genome projects, including our own. Responsible for developing several techniques critical to speeding genome sequencing efforts, Venter is not known for his modesty. Said he in 1999 of that first accomplishment, "The sequencing of the *Haemophilus influenzae* genome in

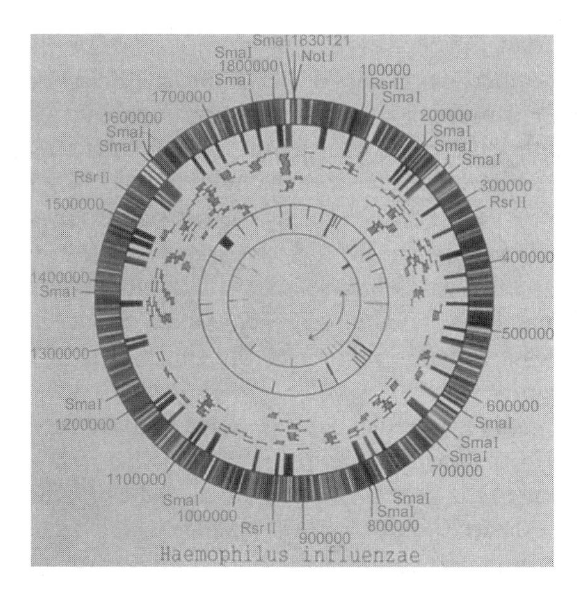

FIGURE 1.3

The complete genome sequence of the bacterium *Hemophilus influenzae*.
Courtesy of and used with permission from Dr. Clare M. Fraser and The Institute for Genomic Research.

1995 will be a bright line in the history of science; events will be described as occurring before or after that. The human genome sequence will be one more blip along the way. We are coming out of the dark ages and into the enlightenment." Yet antagonism toward him has diminished somewhat as his contributions continue to mount.

The sequencing of *H. influenzae* was followed in rapid order by the unveiling of other genomes, chosen for their diversity or pivotal place in evolution. Not all have sprung from Venter's companies. The ever-growing list includes the smallest organism's "near-minimal set of genes for independent life" (*Mycoplasma genitalium*); the first member of the Archaea domain (*Methanococcus jannaschii;* see Chapter 3); the laboratory standard *E. coli;* a hardy bacterium that can live in nuclear reactors called *Deinococcus radiodurans;* the yeast *Saccharomyces cerevisiae;* a simple multicellular organism, the roundworm *Caenorhabditis elegans*; rice; the fruit fly, *Drosophila melanogaster*, and of course, *Homo sapiens*, unveiled in June 2000.

Many biologists eagerly await the sequencing of organisms more connected to ourselves. "It will be fascinating to sequence the genomes of our closest primate relatives. A chimp's genome differs from a human's by only one percent, and all our differences are buried there. The chimp probably has every gene we do, but the differences lie in gene control and regulation," says Leroy Hood, a professor of genetics at the University of Washington in Seattle who contributed to the rice genome project and, in the 1980s, invented the automated DNA sequencers that evolved into the instruments used today.

Human pathogens are understandably high on the priority list for genome sequencing. Among the first sequenced were the bacteria that cause syphilis, Lyme disease, and tuberculosis, and the organism that causes malaria. *Helicobacter pylori*, the main ingredient in Barry Marshall's "swamp water" and the cause of peptic ulcers, has also had its genome sequenced.

Once many larger genome sequences become available, researchers will be able to apply information from one species to others, revealing and taking advantage of evolutionary relationships. Consider the potential benefits to agriculture. Corresponding genes in closely related crop species are generally so similar in sequence that a gene from one species can substitute for its counterpart in another. "All cereals are members of the grass family, and they have a lot of similarity in their origin. All of the technology developed in one cereal will work for all cereals. Now when we want to improve a crop, we don't have to go through 50 years and 500 careers to do it," says John Axtell, a distinguished professor of agronomy at Purdue University. For example, the corn genome—about 2 billion DNA bases—is about three times the size of that of sorghum. "Today, the fastest way to improve corn might be to study the genetics of sorghum. By identifying genes for a desired trait, such as drought tolerance, in sorghum, researchers know where to look for it in corn," Axtell adds.

Sequencing genomes has two types of aims: providing knowledge about different species, and improving the human condition through more directed approaches, such as singling out disease-causing organisms. "Whole genome analysis so far has been a fairly skewed sample. I worry that it will be difficult to study biodiversity because even though the cost of sequencing has gone down, it is still expensive. NIH [National Institutes of Health] focuses on pathogenic organisms, and DOE [Department of Energy] on those of industrial relevance. So much of the microbial world may remain unexplored," says Clare Fraser, president of the Institute for Genomic Research in Rockville, a facility that has sequenced many genomes.

The genome sequences revealed so far indicate how very much we do not know, despite decades of cataloging individual genes. "Each genome exhibits a different theme, which argues against the closed mindset that once said all we'd need to understand the microbial universe was to know *E. coli*. It is amazing how much new biology is coming out of understanding genomes," Venter says. After the first sequenced genomes revealed that up to one-third of their genes had no known counterparts in other species, researchers thought that this high number of unknowns merely reflected the fact that it was early in the game. These were just the first puzzle pieces. But as databases of other genomes accrued, many parts of many genomes remained mysteries. Even *E. coli*, the best-studied microorganism because of its critical role in the last century in working out the details of protein synthesis, has not revealed all its secrets—we know the functions of only 2000 of its 4288 protein-encoding genes. Comparisons can be confusing too. For example, the fruit fly has fewer protein-encoding genes than the much simpler roundworm *C. elegans*, and only about twice as many genes as yeast! Clearly, gene number alone does not correlate to complexity.

It indeed seems that the more we learn about genomes, the more we realize there is yet to discover—something that can be said about science in general.

THE HUMAN GENOME PROJECT

It was perhaps inevitable that discovery of the DNA double helix in 1953, and the ensuing decades of deciphering how genes orchestrate cellular activities, would culminate with an attempt to determine the sequence of the entire human genome (Table 1.3). Like climbing the highest mountain or traveling to the moon, the ability to sequence genes that became possible in the 1980s begged the larger question: Could we sequence our own genome? "It was an idea whose time had come. A whole series of people were involved.

TABLE 1.3

Evolution of the Human Genome Project	
1985–1988	Idea crystallizes at several scientific meetings.
1988	Congress authorizes DOE and NIH to fund project.
1990	Project begins.
1991	Invention of expressed sequence tag (EST) technology.
1994	France and United States publish map of 6000 markers, 1 million bases apart.
1995	Shift in focus of project from gene mapping to gene sequencing.
1996	Resolution passed at Second International Strategy Meeting on Human Genome Sequencing to make all data available freely in the public domain. Updated daily at GenBank (http://www.ncbi.nlm.nih.gov).
1998	Public genome sequencing consortium publishes preliminary map of pieces covering 98% of genome. Deadline moved up to 2003.
2000	Celera Genomics Corporation announces "first draft" of human genome sequence, based on whole genome shotgunning.
2002	Anticipated date of public consortium human genome sequence, based on incremental approach.

No one person can claim he or she was the author of the idea; it was in the air," recalls Leroy Hood.

In 1984, the idea of sequencing the human genome was raised at a Department of Energy–sponsored conference as a way to monitor biological response to radiation exposure, a matter of great concern at that time because prominent scientists were evoking "nuclear winter" scenarios that might follow the use of nuclear weapons. Hood recalls that in the spring of 1985, Robert Sinshimer, chancellor at the University of California, Santa Cruz, also proposed establishing an institute to sequence the human genome at a brief meeting of many of the leaders in genetics. "It was a defining moment. I had thought it was an exciting idea, but technically impossible in terms of how much there was to do." On the East Coast in 1986, renowned virologist Renato Dulbecco suggested that knowing the genome sequence would reveal how cancer arises. That summer the idea for a genome project further crystallized as geneticists met at the Cold Spring Harbor Laboratory on New York's Long Island.

But not everyone embraced the idea. Debate raged in scientific circles. The primary intellectual objection echoes the dichotomy between the artistic and the systematic scientist. Recalls Hood, "A determined majority fought the start of the genome project on the grounds that it wasn't good science—it was not discovery or hypothesis driven, more like stamp collecting. The project was, instead, a matter of seeing what's there and creating raw materials and tools to do biology in a different way. The idea was revolutionary." A practical objection was the apprehension that a government-funded human genome effort would divert funds from other research areas, a critical concern at the height of the AIDS pandemic. To settle the issue of whether to move forward, the National Academy of Sciences convened a committee of scientists to discuss the feasibility, risks, and benefits of attempting to sequence the human genome. The group started out composed of one-half supporters and one-half detractors, but after much discussion, Hood remembers, all agreed it was a great idea. In 1988, Congress authorized the National Institutes of Health and the Department of Energy to fund a $3 billion, 15-year human genome project. For the first years, the Department of Energy was the more active participant.

And so in 1990 started what Francis Collins, director of the U.S. National Human Genome Research Institute, calls the deciphering of "the molecular essence of humankind." From the outset, 5% of the federally funded budget was set aside to fund the "ELSI" (Ethical, Legal, and Scientific Issues) program to help ensure that genetic information, as it is uncovered, is not used to discriminate against individuals. The public sequencing effort has gone global, with 10 major sequencing centers, including facilities in England, France, Germany, and Japan. Many hundreds of other laboratories are also contributing sequence data, as are companies. Although the human genome project is an organized, "big science" effort, it builds on the work of many individual researchers who have been studying, mapping, and sequencing genes for many years. There was, indeed, a robust field of genetics that predated the genome project, as well as the modern biotechnology field that began in the mid-1970s.

The human genome project has evolved in stages. For the first few years, the focus was on developing the tools with which to cut the genome into workable-sized pieces and on refining the computer capability to overlap pieces to derive continuous sequences of DNA bases. Researchers assembled preliminary maps, using known genes as guideposts among the 24 chromosome types, like indicating state names on a map of the United States. Much of those initial data came from existing studies on disease-causing genes within families, some dating back decades. And each October, in synchronization with the annual meeting of the American Society of Human Genetics where researchers report newly sequenced genes, *Science* magazine charted the project's progress with spectacular chromosome maps that became ever more crowded with landmarks each year. It's been exciting to see the maps fill in, a little like progressing from the outline of the 50 states to adding cities, then towns, and finally roads. An alternate approach, taken by Celera and others, was to "shotgun" the entire genome into pieces, which a supercomputer then overlapped to derive the sequence.

During the "middle years" of 1993 through 1998, technology improved, with sequencing rate accelerating by an order of magnitude. At the same time, researchers began to focus on genes that might be of clinical importance by using expressed sequence tags (ESTs), an approach pioneered by Craig Venter in 1991. ESTs are pieces of DNA that represent genes known to encode proteins—about a mere 5% of the genome. They are derived from the genes that are expressed in a particular cell type. For example, analysis of the ESTs unique to bone cells might shed light on the development of bone diseases such as osteoporosis or rickets. With 3 billion bases of human DNA to wade through, EST technology cuts through the less interesting stuff—a little like using a search function to rapidly locate information in a database.

The final years of the human genome project unfortunately were and continue to be intensely competitive, although the public and private efforts came to a last-minute agreement to announcedual success together, on June 26, 2000. The *Drosophila* genome project, in contrast, was throughout a collaboration between a government-funded consortium and Celera Genomics, culminating in an "annotation jamboree" in late 1999 where 300 researchers gathered to assign gene functions to DNA sequences. Looking beyond the sequencing, the field of genomics has emerged, which considers gene interactions. It's a whole new arena for discovery. Genes control each other and respond to environmental cues, and deciphering these hierarchies of expression will be no small task.

Genome researchers aren't daunted by how much remains to be discovered and understood, a challenge that will probably take longer than their lifetimes. "We are realizing that what biology is about is complex biological systems and networks. We will be able to use genome information to understand how the brain works, how immunity works. System properties arise from the interactions of several genes, several proteins. This is a fundamental paradigm change in the way we see biology," relates Hood.

After the Genome Project Just as technology made possible the sequencing of genomes, so too is it fueling the next step, the analysis of multiple genes. But unlike the armies of bulky sequencers that crunched out strings of A, C, T, and G, the device that is transforming genetics to genomics is tiny—a mere chip.

A DNA chip is a solid support, typically made of nylon or glass, that contains carefully chosen short pieces of DNA. (It is also called a *DNA microarray.*) To the chip is added DNA from a sample—a bit of tissue from a cancerous tumor, for example. The sample DNA is labeled so that it gives off a fluorescent signal when it binds to its complement on the chip. The resulting pattern of glowing dots on a DNA chip can provide a wealth of information. One such device, called a "lymphochip," is embedded with several thousand genes that, when mutant, contribute to the development of lymphoma. An individual's pattern of fluorescing genes can tell a physician which subtype of cancer he or she has. Another type of chip harbors genes that reveal whether that particular patient's cancer cells will take up and metabolize a particular drug, enabling a physician to determine, prior to treatment, which drugs are likely to work the best. And as a cancer changes over time, different types of chips can be tailored to monitor progress. DNA chips will also prove invaluable in evaluating and treating infectious diseases. Infectious disease chips will be used to identify specific pathogens, predict and select effective treatments, and monitor a patient's response. With the advent of DNA chips, genetics may progress from the systematic science of genome sequencing to the more artistic challenge of choosing meaningful gene combinations.

Yet even with the promise of DNA chip technology to radically individualize the practice of medicine, life in the postgenome era raises fears of too much being known about people. Will the day come when genome sequencing is as routine a procedure at a birth as pricking a newborn's heel to test for a half-dozen inherited diseases is today? Who will have access to a person's genome information? A science fiction film called GATTACA dramatized life in a highly restrictive futuristic society where the government knows everyone's genome sequence. Efforts by several countries to create nationwide DNA databases—led by Iceland but also including the United Kingdom, Estonia, and Sweden—have galvanized concerns that GATTACA may one day be real. Although the stated immediate goal of such databases is to amass medical information that can improve health care, details of precisely how that information is to be used, and by whom, have yet to be worked out.

What will discovery be like in this new century? Will the highly organized projects of big science—the combinatorial chemistries and genome projects and who knows what else—sap the creativity from research or open up vast new territories to explore, with as yet unimaginable tools and technologies? Will public funding and industry interest come to favor directed research, with specific human-centric goals, as opposed to the basic research that seeks answers to fundamental questions about how nature works? Will there still be room for serendipity? Will there be a place for the lone but driven scientist who has the nerve to follow an unusual idea or who sees connections that elude others? Let's hope that the future continues to have a place for both the artistic and the systematic scientist.

The following chapters chronicle several major discoveries in the life sciences, some basic enough to interest primarily biologists, such as origin-of-life research, and others that have captured the imaginations of nearly everyone, such as cloning. What these stories of life share are the initiative of the discoverers to notice the unusual, follow hunches, take risks, and, finally, assemble the pieces of a compelling intellectual puzzle.

REFERENCES

Annas, G. J. "Rules for research on human genetic variation-lessons from Iceland," *The New England Journal of Medicine* 342:1830–1834 (2000).

Lewis, Ricki. "From basic research to cancer drug: The story of cisplatin." *The Scientist,* 5 July 1999, 8.

Lewis, Ricki. "Iceland's public supports database, but scientists object." *The Scientist,* 19 July 1999, 1.

Lewis, Ricki. "Confessions of an ex fly pusher." *The Scientist,* 1 May 2000, 1.

Lewis, Ricki. "Evolving human genetics textbook chronicles field." *The Scientist,* 24 July 2000, p. 34.

Marshall, Eliot. "A high stakes gamble on genome sequencing." *Science* 284:1906–1909 (1999).

Rose, Peter G., et al. "Concurrent cisplatin-based radiotherapy and chemotherapy for locally advanced cervical cancer." *The New England Journal of Medicine* 340:1144–1153 (1999).

Rosenberg, Barnett. "Platinum complexes for the treatment of cancer." *Interdisciplinary Science Reviews* 3:134–147 (1978).

Rothwell, Nancy J. "Show them how it's *really* done." *Nature* 405:621 (2000).

Schou, Mogens. "Lithium perspectives." *Neuropsychobiology* 10:7–12 (1983).

Schou, Mogens, et al. "The treatment of manic psychoses by the administration of lithium salts." *Journal of Neurological Neurosurgical Psychiatry* 17:250–257 (1954).

Service, Robert F. "Race for molecular summits." *Science* 285:184–187 (1999).

Watson, James. "The double helix revisited." *Time,* 3 July 2000, p. 30.

CHAPTER 2

THE ORIGIN OF LIFE: WHEN CHEMISTRY BECAME BIOLOGY

It's nearly impossible to imagine our planet without its ever-changing coating of life, but for several hundred million years, the earth lacked life of any kind. Then long ago, a collection of chemicals came together that was unlike others before it. The collection formed a system that could use energy from the environment to continue existing and possibly even grow, and that could reproduce. It was alive.

Perhaps these chemical assemblies appeared many times, in different ways or with different components, with only some, or one, surviving. We cannot know precisely what happened as life debuted because we cannot turn back the clock. But we can ask questions about the events that transformed chemistry into biology, and devise experiments to investigate how the chemicals found in life might have formed and begun to interact and aggregate in an organized manner.

The experiments that attempt to retrace the steps preceding life's beginnings cap centuries of inquiry that sought to answer the question of how life originated and originates. But these early investigations often dodged the central question.

SIDESTEPPING THE QUESTION OF HOW LIFE BEGAN

Life on earth is remarkably diverse, and has been for quite some time. Organisms range from complex billion- and trillion-celled plants, fungi, and animals to a wealth of single-celled organisms whose diversity we can barely imagine, let alone measure. Scientific explanations of how life began follow the logic of increasing complexity, assuming that a microbe similar to modern-day bacteria probably lived before more complex single cells such as paramecia. Likewise, single-celled life preceded multicelled life. In contrast, non-scientific explanations for life's origin envision complex organisms appearing, fully formed, on the earth. The first life mentioned in Genesis, for example, is grass—a flowering plant and a complex organism.

The religious approach does not answer the question of how life began, of *how* the Creator created. Explaining an event without invoking a mechanism is not science.

DEBUNKING SPONTANEOUS GENERATION

For many years, people, including scientists, thought that life could "spontaneously generate" from nonliving matter. This idea persisted for centuries because it fit everyday observations. For example, a puddle that was apparently lifeless one day would contain worms the next. Similarly, mice and cockroaches could appear as if from nowhere in the walls of a house.

It took several series of experiments to disprove spontaneous generation. In 1668, Italian physician Francesco Redi put meat into eight containers, leaving four open to the air and sealing four shut. Maggots appeared only in the open flasks, where flies could enter. Did the fresh air bring the maggots? To answer that question, Redi closed four of the flasks with gauze, allowing fresh air in but keeping flies out. No maggots appeared. Redi concluded that adult insects were necessary to generate maggots, but his results weren't accepted for very long. When microscopes revealed microbial life, people wondered anew about spontaneous generation. Could bacteria appear from nothingness?

English naturalist John Needham launched a new attack on spontaneous generation, but at first he thought that his experiments actually supported the theory. In 1748, he boiled mutton broth for a few minutes in sealed glass vessels, and when bacteria appeared, he thought that they did so spontaneously. Two decades later, Italian biologist Lazzaro Spallanzani claimed that Needham hadn't boiled his soup long enough to kill bacterial spores. Finally, by boiling longer and sealing the flasks, Needham kept bacteria from growing, disproving spontaneous generation.

But many people, including some scientists, still weren't convinced, objecting that the boiling killed an undefined "vital principle" necessary for life. It would be another century before yet another investigator, Louis Pasteur, would definitively disprove spontaneous generation by boiling meat broth in flasks that had S-shaped curved open necks. Air could get in, but not the dust particles that carry bacterial spores. Because he didn't use corks, gauze, or glass seals, Pasteur's flasks didn't block the vital principle. Yet still no microbes grew. After a public announcement and evaluation by a group of luminaries of the day, it was decided that Pasteur had finally disproved spontaneous generation.

Disproving spontaneous generation did not answer the question of how life began, however. It only showed that life comes from preexisting life.

DISPROVING VITALISM

Vitalism held that different laws than those underlying chemistry and physics controlled living matter, and this proclamation was enough to stop some people from questioning further. But experiments soon showed that living matter is indeed subject to the laws of nature. Throughout the nineteenth century, chemists and physiologists began to chip away at the vague and all-encompassing veil of vitalism.

In 1828, German chemist Friedrich Wohler challenged the prevailing idea that chemicals that are part of an organism came from a "vital force." He synthesized urea, a waste product of metabolism. If urea was unique to living organisms, Wohler reasoned, a chemist wouldn't be able to make it in a test tube. A generation later, French chemist Pierre Eugene Berthelot synthesized many carbon-containing chemicals, some identical to those in organisms and some not. His work led to redefining *organic* to mean a compound containing carbon, rather than a carbon-containing compound found in life.

The early physiologists also took a rational approach to replacing vitalism with demonstrations of the physical causes of biological functions. German physiologist Karl Friedrich Wilhelm Ludwig created a device that measured blood pressure, revealing that this is the force that propels blood. German physiologist Emil Heinrich DuBois-Reymon showed in fish that nerve impulses aren't magical, but electrical. Another German physiologist, Max Rubner, demonstrated that the energy of life ultimately comes from an animal's food, which a body burns much as a fire consumes objects. Then German chemist Eduard Buchner showed that juice squeezed from killed yeast cells could still cause carbon dioxide to bubble up from sugar. How could this reaction of life—fermentation—continue if the vital force had clearly been snuffed out of the yeast? Ultimately, experiments buried vitalism.

But disproving the input of a vital force did not answer the question of how life began. It only showed how life did not begin.

LIFE FROM SPACE: PERHAPS SOME EVIDENCE

Could life have come here from beyond the earth? Although most scientists acknowledge the possibility, if not the certainty, of extraterrestrial life, the hypothesis that it is the source of the origin of life here is less accepted. But the idea has definitely generated much thought.

Swedish physical chemist Svante Arrhenius suggested that life arrived from space in his 1908 book *Worlds in the Making.* Arrhenius was no stranger to unusual hypotheses—he nearly flunked out of graduate school for his theory of ions (charged atoms), for which he much later received a Nobel Prize. In his later years, Arrhenius' interests broadened to encompass "panspermia," the idea that interstellar dust, comets, asteroids, and meteorites seeded the earth with spores that survived the cold vacuum of space, driven by the energy of radiation.

Panspermia envisioned entire cells coming to Earth. In the late 1970s, British astronomers Sir Fred Hoyle and Chandra Wickramasinghe amended the idea, suggesting that "traces of life," in the form of biochemical instructions, were the emissaries from the cosmos necessary to spark life here.

The reasoning behind panspermia is that the earth isn't old enough to have accrued the diversity of life on the planet today. Jump-starting life with cells or biochemical blueprints from beyond the earth, however, could account for the millions of species. But data supporting an extraterrestrial origin of life were sorely lacking, and the very idea evoked images of science fiction more than it did science because panspermia seemed untestable, at least directly. But over the past few years, a convergence of intriguing and controversial discoveries has lent possible support to the concept of life from space, spurring reemergence of the idea of panspermia under the heading "cosmic ancestry."

Life from Mars? One discovery that has revitalized panspermia is a grapefruit-sized chunk of rock found in Antarctica on December 27, 1984. The meteorite, known officially as Alan Hills 84001 or simply ALH84001, landed on Earth some 13,000 years previously, having formed on Mars about 4.5 billion years ago (Figure 2.1). But the event that interests origin-of-life researchers probably occurred 4 to 3.6 billion years ago, when liquid water, likely at a temperature that can support life, percolated through cracks in the rock. At these sites tiny lumps of organic materials called polycyclic aromatic hydrocarbons (PAHs) formed. A team of scientists led by David McKay of the Lyndon B. Johnson Space Center in Houston dissolved these PAHs in acetic acid, releasing grains of the mineral magnetite that looked like hexagons when viewed from a certain angle. The neat rows of hexagons resembled remains of a certain type of bacterium, but did not look at all like magnetite precipitated from nonliving materials.

When *Science* magazine published an article late in the summer of 1996 entitled "Search for Past Life on Mars: Possible Relic Biogenic Activity in Martian Meteorite ALH84001," by McKay's team, it caused such a ruckus that President Clinton commented on the work at a press conference. His sound bites were later, without permission, spliced into *Contact,* the film based on Carl Sagan's novel about a rendezvous with extraterrestrial beings. Despite the ensuing media hype concerning possible life on Mars, the scientific community remained skeptical, as scientists tend to do. For example, Jeffrey Bada, a professor

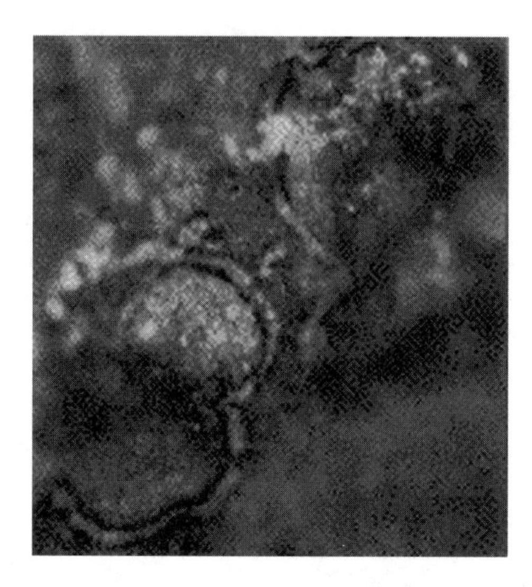

FIGURE 2.1

ALH84001 is a meteorite of Martian origin. The light areas on the Martian meteorite are grains of carbonate, indicating immersion in water. Within the carbonate, researchers found magnetite inclusions that resemble terrestrial bacteria. Are these organic inclusions traces of bacteria-like cells?
Courtesy of NASA.

of marine chemistry at the Scripps Institution of Oceanography in San Diego, argued against the carbonate globules indicating life. He pointed out that a book published in 1888 called *Meteorites and Their Origins* made similar claims, as had reports in the 1960s of "organized elements" in meteorites, which had turned out to be degraded ragweed pollen.

A major criticism of the Mars meteorite work was that the bacteria envisioned to have left the PAHs would have been far smaller than their terrestrial counterparts. But McKay and his team are keeping open minds. "Where is it written that Martian bugs have to be identical to Earth bugs?" McKay asks. Researchers on both sides of the controversy agree that the only way to settle the question is to obtain more evidence. Ironically, some of that evidence was already at hand two years after ALH84001 made its prominent debut in *Science*. In May 1998, the Mars Global Surveyor detected a 300-mile stretch of hematite on the Red Planet, a mineral that forms in the presence of warm water. Phil Christensen, the principal investigator for the Surveyor project at Arizona State University, says that the finding provides "the first evidence that suggests a large-scale hydrothermal system may have operated beneath the Martian surface at some time during the planet's history. This is one of the best places to look for evidence of life on Mars." Might Martian life have seeded the earth?

Unfortunately, more recent attempts to probe the Red Planet's surface have been less than successful, if not downright embarrassing. In September 1999, the Mars Climate Orbiter was lost when operators at one center in Denver reported information in units of pound-seconds, whereas researchers at the receiving end, at the Jet Propulsion Laboratory in Pasadena, assumed that the data were being sent in units of newton-seconds. After that fiasco, extra measures were taken to improve the odds of success for the Mars

Polar Lander, set for Martian arrival on December 3, 1999, at a price tag of $165 million. It reached Mars, but hasn't been heard from since. Cost-cutting measures had silenced the device as it traversed the Martian atmosphere, slowed by parachute, and finally landed with the aid of rocket boosters.

Clues in Chemical Handedness Shortly after ALH84001 refocused attention on panspermia in the guise of cosmic ancestry, another meteorite reappeared on the scene of origin-of-life research. In 1997, John Cronin and Sandra Pizzarello of Arizona State University discovered mostly left-handed forms of an amino acid in samples of powder from the Murchison meteorite, a 100-kilogram rock that fell in Australia on September 29, 1969, and has been probed by researchers ever since. This was odd, because amino acids (which link to form proteins) and some other chemicals are known to exist in two forms that are mirror images of each other, much like a left hand and a right hand, which cannot be superimposed. Such molecules are termed *chiral,* which is Greek for "hand."

In a laboratory-synthesized version of a chiral molecule, the left-handed and right-handed forms are present in roughly equal amounts. But when the same type of molecule occurs in an organism, it is present in only one form. Amino acids in life are always left-handed, and the sugars in the nucleic acids RNA and DNA are always right-handed. Vitamin C is always right-handed, as is the sugar glucose. Certain drugs only work if they are of one handedness.

Louis Pasteur discovered chirality when he was just 25 years old. As a chemist, he was intrigued by racemic acid and tartaric acid. They were structurally the same, but differed in their ability to rotate polarized light. Normally light consists of waves traveling at many angles; polarized light consists of waves aligned in a single plane. Racemic acid could not rotate the plane of polarized light, but tartaric acid could. Pasteur decided to investigate why this happened by looking at crystals of both types of acids. Using a hand lens, he could see that racemic acid actually consisted of two types of crystals, and what's more, they appeared to be mirror images of each other. One type rotated the light in one direction, and the other type in the other direction, cancelling each other out so that there was no effect on the light. But tartaric acid consisted of only right-handed crystals, and these rotated the plane of polarized light. Tartaric acid, it turned out, was actually the right-handed component of racemic acid. The left-handed form of tartaric acid was an entirely new compound! Over the years, Pasteur as well as other biologists discovered many compounds that can assume two chiral forms, but are present as only one in organisms. Such one-way molecules are very much a fact of life.

Different scenarios may explain why certain biochemicals exist in organisms in only one chiral form. If life began on Earth, its trademark handedness may have simply been due to chance, with the first cell incorporating a left-handed amino acid and a right-handed nucleic acid precursor. Or perhaps life on Earth began with two types of amino acids or nucleic acids, or both, and one type came to persist because it offered some advantage. Alternatively, maybe the chirality question predates the origin of life on Earth: Could one chiral type have arrived on Earth on the celestial vehicles of panspermia? The finding of mostly left-handed amino acids in the Murchison meteorite suggests that this might be the case.

In 1998, Jeremy Bailey at the Anglo-Australian Observatory near Sydney discovered how single-form amino acids might have fallen to Earth. Using a special camera, he examined infrared light coming from a part of the Orion Nebula where new stars arise, a region thought to be similar to the early solar system. Much of the light was rotating in a single direction, a phenomenon called circular polarization that occurs when light scatters off dust particles that are aligned in a magnetic field. Previous experiments had shown that circularly polarized light can selectively destroy one form of a chiral molecule. Therefore, finding such light from space suggests a way that predominantly left-handed amino acids might have arrived on the earth—they could have selectively survived a run-in with magnetized dust.

Life in Extreme Environments More down-to-earth evidence cited to support cosmic ancestry comes from *extremophiles*—microorganisms that live in extreme (to us) habitats. Charles Darwin envisioned life brewing into existence in a "warm little pond." A more likely scenario based on geological clues includes volcanoes, earthquakes, lightning, and an atmosphere that allowed ultraviolet radiation to bathe the planet. The conditions were probably similar to those on the outer planets today—hence the idea that if life started here under such harsh conditions, it might have come from similar environments elsewhere.

For some organisms, the high radiation or heat of the early earth might not have been a deterrent. Clues from present-day life include a large, reddish bacterium called *Deinococcus radiodurans* discovered in a can of spoiled ground meat at the Oregon Agricultural Experiment Station in Corvallis in 1956. It apparently withstood the radiation used to sterilize the food, and has since been found to tolerate 1000 times the radiation level that a person can. It even lives in nuclear reactors! When radiation shatters the bacterium's genetic material, it stitches the pieces back together. *The Guinness Book of Records* dubbed *D. radiodurans* "the world's toughest bacterium." Its genome, sequenced in late 1999, may reveal how it survives in the presence of intense radiation.

The radiation-resistant giant bacterium is not alone in being able to live under early Earth conditions. In the 1960s, Thomas Brock, a microbiologist then at Indiana University, dipped glass microscope slides into near-boiling hot springs in Yellowstone Park. Days later, the slides sported bacterial colonies so robust that they were visible to the naked eye. Brock called the microbes "relicts of primordial forms of life," writing in *Science* magazine that "it is thus impossible to conclude that there is any upper temperature of life." But the Yellowstone hot springs inhabitants look like cool customers compared with *Pyrolobus fumarii* bacteria. These microorganisms thrive at 113° Celsius (235° Fahrenheit) in chimney-like "black smokers" that rise from the Pacific seafloor in areas called deep-sea hydrothermal vents, which spew superheated water tinged with sulfurous chemicals from the hot mantle beneath. These bacteria and others live under great heat and pressure. They extract energy from the sulfur-based compounds in the environment, supporting a community of huge worms, sea spiders, limpets, and other organisms.

Other present-day extremophiles suggest different panspermia origins. Researchers from Montana State University have discovered microbial "oases" in pockets of liquid water that form in midsummer within the permanent ice layers of lakes in Antarctica's McMurdo Dry Valleys. If life can colonize frozen lakes, might it also have survived in the icy interior of a comet headed for Earth? We also know that microorganisms can live in rocks. These rock dwellers, called *cryptoendoliths*, live 500 meters (1640 feet) below the surface of the Savannah River in Aiken, South Carolina; in volcanic basalt 2800 meters (1.33 miles) beneath the surface of the Columbia River in Richland, Washington; at the boundary between shale and sandstone in Cerro Negro, New Mexico; and in water-filled cracks in granite 1240 meters (4067 feet) beneath the earth's surface in Manitoba and Sweden. If life can colonize rocks, might it also have survived aboard meteors headed for Earth?

Apparently, life as we know it can find nourishment, energy, and space in places we never thought it could. But finding bacteria-like structures and the one-sided chemicals of life in meteorites, evidence of water on ancient Mars, and life on Earth today that could have survived the harsh conditions of another world and another time does not answer the question of how life began *here.*

SETTING THE SCENE

When chemistry became biology, the earth was not like it is now. Although we don't have any direct evidence of what the conditions were, clues from astronomy and geology paint a portrait that differs considerably from Darwin's warm little pond.

WHEN MIGHT LIFE HAVE ORIGINATED?

Astronomy tells us that the earth formed about 4.55 billion years ago, cooling sufficiently to form a crust by 4.2 to 4.1 billion years ago. Surface temperatures then ranged from 500° to 1000° Celsius (932° to 1832° Fahrenheit), with atmospheric pressure 10 times what it is now. Little if any of this ancient crust remains. It has been crumbled and rebuilt into sediments that have long since been altered by heat and pressure or been sucked down into the earth's interior and possibly recycled to the surface.

Craters on the moon are testament to an asteroid bombardment there from 4.0 to 3.8 billion years ago. If those asteroids also hit the earth, they would have wiped out any organic aggregates that might have induced the beginnings of life. But Gustaf Arrhenius of the Scripps Institution of Oceanography (and grandson of Svante Arrhenius of panspermia fame) says that evidence for such an earthly bombardment isn't clear, and that the oldest known rocks, from Greenland, appear to have escaped this fate. "The Greenland rocks were already there, 500 to 600 million years after the formation of the earth," he says. And also there, he thinks, was life.

Arrhenius' group discovered the earliest known traces of life in those rocks, in an iron-rich band of material called the Isua geological formation. These clues aren't squiggly worms or feathery ferns embedded in rock, but a chemical signature that indicates cells may have once been there. To discover these telltale signs of the once living, the researchers probed quartz that contains organic deposits that have an excess of a lighter form of carbon known to come from the metabolism of a living organism (Figure 2.2). A purely physical process couldn't account for the ratio of light to heavy carbon in these ancient rocks. Some of the oldest actual fossils are from 3.7-billion-year-old rock in Warrawoona, Australia, and Swaziland, South Africa. These microfossils are minerals that replaced cells, which aggregated into large formations called stromatolites.

If the earth cooled by about 4.2 billion years ago, and the first traces of biochemistry we know of date to 3.85 billion years ago, that leaves about 350 million years for life's precursors to have formed. Chemists say that is enough time. But what was the place—or places—where it happened?

WHERE COULD LIFE HAVE ARISEN?

Since the earth was still in geologic chaos around 4 billion years ago, the chemicals that led to life probably formed, accumulated, and interacted in protected pockets of the environment. Deep-sea hydrothermal vents, where the earth's interior bubbles onto the

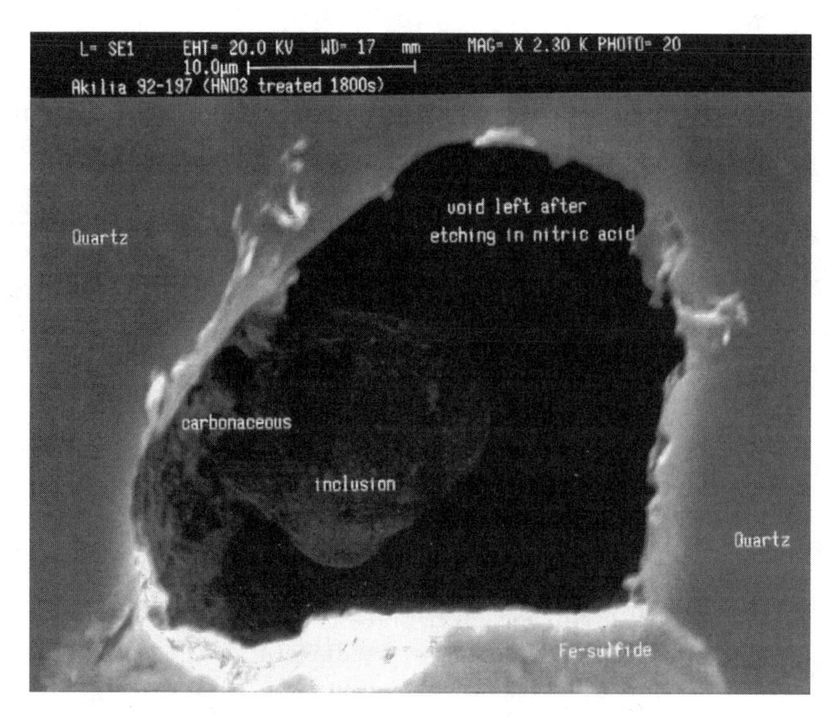

FIGURE 2.2

The ratio of carbon isotopes in organic inclusions in ancient rocks may be a vestige of long-ago life.
Courtesy of Gustaf Arrheniu, Scripps Institute of Oceanography, University of California, San Diego. Scanning electron micrograph by S. Mojzsis.

ocean floor, are a possibility. A real estate agent hawking these areas as choice property for the start of life would mention several selling points. Their location could have shielded the delicate organic chemical collections from the wrath of earthquakes and volcanoes. The warmth they provided to surrounding areas (away from the furnacelike conditions at the vents themselves) would have fostered chemical reactions. Perhaps most important, the rich minerals spewing from the mantle may have functioned as catalysts and served as a physical mold on which chemicals could have linked to build larger molecules.

The deep-sea hydrothermal vents of today are holdovers, perhaps, of environments that were more common on the much hotter early earth, but they cannot harbor direct descendants of those first organisms. "All environments on earth are 'contaminated' with modern life, from the deep continental bedrock, to high temperature hydrothermal vent chimneys. While we can learn much about how modern life has adapted to these environments, any remnants of the first life forms were surely wiped out long ago by competition," says Jay Brandes, a researcher at the University of Texas who builds laboratory models of vents. Still, studying the living communities that border deep-sea hydrothermal vents can suggest ways that life might have originated.

Whenever and wherever the transition from chemistry to biology occurred, it may have been fast—by geological standards, anyway. Stanley Miller, professor of chemistry at the University of California at San Diego, thinks life had to have begun within a 10-million-year time frame because that is the time it takes for the entire ocean to recirculate through vent systems. Once sucked down into the vents, heat would have torn apart the delicate molecules of life. "The origin of life probably occurred relatively quickly. It could have gone from the beginning 'soup' to cyanobacteria [blue-green algae] in 10 million years. We used to talk of billions of years, but in the origin of life, time is a detriment because compounds decompose," Miller says. Volcanoes, earthquakes, and debris from space would also have dismantled organic molecules before they could associate into anything approaching a biologically useful molecule.

PREBIOTIC SIMULATIONS

The fact that we cannot travel back in time, and that modern life "contaminates" relict areas where life might have arisen long ago, has not deterred scientists keen on discovering how life began. Designing experiments that mimic geophysical conditions on the early earth, and combining chemicals that could have existed then, is the essence of origin-of-life research. "We have absolutely no idea about how easy or difficult it was to make life. The way to approach this question, though, is to model experiments in the lab to try to find processes that could have, in nature, spontaneously given rise to fundamental information, particularly RNA," says Gustaf Arrhenius.

The era of prebiotic simulations—often couched in gastronomic terms such as "primordial soup" or "mineral crepes"—began nearly half a century ago, with the work of a graduate student, Stanley Miller.

THE MILLER EXPERIMENT

In 1953, Miller built a glass device that contained components of a possible early atmosphere, added a spark, and watched amino acids form. People familiar with biology and chemistry understood the significance of what he had done, but others grossly misinterpreted the elegantly straightforward experiment. "People made jokes. They suggested that I'd grown a rat or a mouse in there!" Miller recalls.

The "Miller experiment" was destined to be repeated in many labs with many variations and to earn a permanent berth in biology textbooks. Yet to this day, Miller is frequently credited with "creating life" in the lab. Consider the great leaps that one newspaper article took 45 years after Miller's experiment: "At least once in the history of the universe, the inanimate materials of a newborn and sterile planet—hydrogen, ammonia and methane—gathered themselves into organisms. And life on Earth began . . . . The quest to recreate the origin of life in a laboratory began in 1953, with a 23-year-old graduate student named Stanley Miller."

Life's origins were probably not that simple, nor Stanley Miller's goals or contributions that dramatic. In September 1951, as a new chemistry graduate student at the University of Chicago, Miller attended a seminar given by a professor in the department, Harold Urey. The young man listened intently as Urey reviewed the work of Soviet chemist Alex I. Oparin. In 1924, Oparin had published a pamphlet suggesting that life arose from simple organic molecules that linked to form polymers. Few people noticed the pamphlet. Then in 1929, English chemist John B. S. Haldane suggested that the early earth had a different atmosphere, an idea that Oparin embraced and extended in a 1938 book, *The Origin of Life.* He proposed that a hydrogen-rich atmosphere was necessary for organic molecules to form and suggested that it might have contained the gases methane (CH_4), ammonia (NH_3), water (H_2O), and hydrogen (H_2), similar to the atmospheres of the

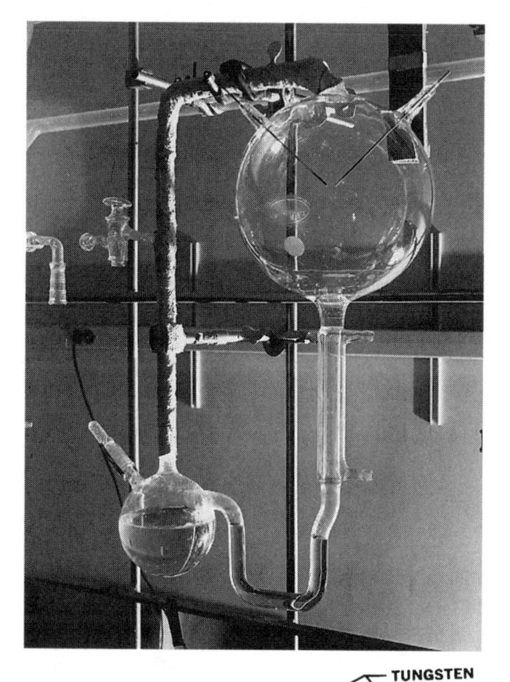

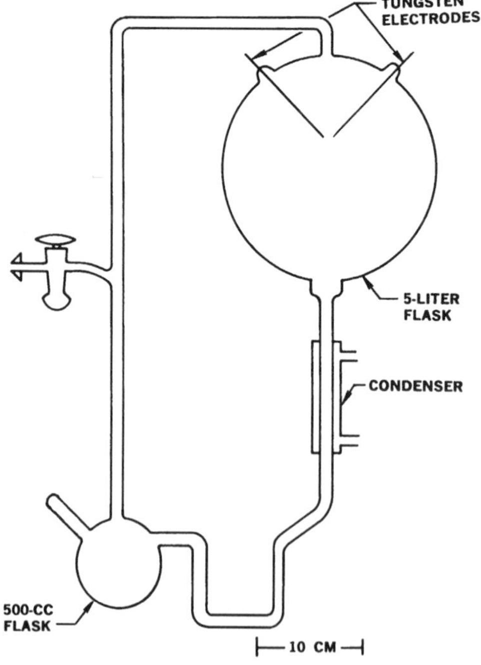

FIGURE 2.3

The apparatus for the Miller experiment modeled an ocean, an atmosphere, lightning, and rain. Methane, water, hydrogen gas, and ammonia in the apparatus reacted to yield amino acids and other simple organic compounds.
Courtesy of Stanley Miller, University of California, San Diego.

outer planets. In contrast, Earth's atmosphere today is rich in carbon dioxide (CO_2), nitrogen (N_2), water, and oxygen (O_2). Why, Urey wondered aloud, had no one ever attempted to experimentally demonstrate what Oparin had hypothesized? Miller wondered too.

After a half-year detour into a theoretical project that avoided the "messiness and time-consuming nature" of experiments, Miller decided to test Oparin's ideas by building an apparatus to house Oparin's four gases. Urey, trying to steer his student on a path toward academic success (that is, experimental results worthy of a PhD), warned him that the experiment might never give him any results.

But Miller accepted the challenge and hit the library stacks to see what others had already done. Several investigators had passed gases through electric discharges to see what compounds would form, and Miller decided that this would be a good stand-in for lightning. He sketched plans for an apparatus and gave them to the university glass shop. The setup would simulate an ocean, atmosphere, and lightning, and used a condenser to make rain (Figure 2.3). Miller and Urey decided to look for amino acids for several reasons: Proteins are important in life, they can be detected easily, and (again being practical) they were likely to form.

In the fall of 1952, Miller filled his glass world with atmosphere, set off the spark, and let it run overnight. The next morning, he was ecstatic to discover a thin layer of material on the surface of the water. He calmed down enough to wait a few more days before removing it, so the deposit could build. Finally, he collected some of the material and placed it on a piece of filter paper and added a solvent. If any amino acids had formed, they would travel along the paper and separate out by size, the largest stopping first. Then he added ninhydrin, a chemical that stains amino acids purple. Nothing happened.

Miller went back to his original plans for the setup, and realized that he had placed the spark and the condenser in the wrong positions—the spark needed to hit the gases before they reached the condenser. So he started again, with a new apparatus. This time, Miller lowered the heat because he feared the modeled world would explode. Two days later, he couldn't detect any material accumulating, but he noticed that the solution had turned yellow. He removed some of the yellow liquid and placed it on the filter paper. When he added ninhydrin, a spot appeared! Its position told Miller that it corresponded to glycine, the simplest of the 20 types of amino acids found in organisms.

On the third try, a braver Miller turned up the heat and waited a week, even though he noticed hues of pink, red, and then yellow-brown the next morning. This time, suspecting that he might have formed more than one type of amino acid, he rotated the filter paper 90 degrees and added a second solvent, to better separate the products. When he added ninhydrin, Miller saw seven purple spots! All were amino acids, although some were not types seen in life. It was a mere 10 weeks since he had started the project.

Miller wrote up what he had done and planned to submit it to the *Journal of the American Chemical Society,* but Urey, recognizing the broader implications of the results, urged his student to send the paper to the prestigious journal *Science.* Urey then took his own name off the paper so that Miller would get the credit. (Had he not done this, the textbooks today would probably describe the "Urey experiment.") While other scientists reviewed the paper before the journal accepted it, Miller repeated the experiment twice more to convince himself of the accuracy of the results. And on May 15, 1953, *Science* published the two-page paper, "A Production of Amino Acids Under Possible Primitive Earth Conditions," by S. L. Miller.

The young graduate student anxiously awaited public reaction, but he was sorely disappointed. Articles in *Time* magazine, the *New York Times,* and then the trickle-down articles in less prominent newspapers and magazines all stressed creating life in the test tube. A Gallup poll even asked if the average person thought it was possible "to create life in a test tube," with 9% responding yes, 78% no, and the rest undecided.

Miller faced scrutiny and skepticism by the scientific community too. When other researchers attempting to repeat his findings found that the red tint he'd seen was not a chemical in life, he demonstrated that it came from an impurity in the glass. When someone suggested to Urey that bacterial contaminants had added the amino acids, Miller repeated the entire experiment yet again, this time baking the device to ensure that all bacteria were kept out, like a modern-day disproving of spontaneous generation. "We sealed and autoclaved the apparatus, and ran it for 18 hours, when 15 minutes would have been sufficient to kill bacteria. And we got the same amino acid products," he recalls.

Miller spent the rest of his time as a graduate student meticulously identifying the chemicals that formed in his apparatus. In the following years, many others built on the idea of a simulated ancient environment. Adding phosphates and ultraviolet irradiation, for example, led to formation of hydrogen cyanide (HCN), an important intermediate for synthesizing amino acids and DNA and RNA bases. Other recipes included carbon dioxide, which may have been present in the early atmosphere. Biochemist Cyril Ponnamperuma produced nucleotides in a Millerlike setup, and noted astronomer Carl Sagan, at the University of Chicago a decade after Miller, synthesized the biological energy molecule and nucleotide ATP (adenosine triphosphate) in a prebiotic simulation.

Then, further confirmation that the chemicals of life could indeed form from chemicals not in life literally dropped from the sky, in the Murchison meteorite. Unlike other meteorites analyzed years after they fell to earth, samples from the Murchison meteorite were studied soon after it fell and contained the same amino acids that Miller had discovered in his experiments, along with others. Miller recalls being "absolutely stunned" at the correspondence between the contents of the rock hunk from space and what had formed in his apparatus. No, he hadn't pulled a rabbit out of a hat (or a rat out of a test tube), but he had elegantly showed that the building blocks of life can form from simpler chemicals and an energy source. It is possible in a lab, and in a meteorite.

Miller is still doing prebiotic simulations. In one recent study, he combined three dry compounds that could have been present on the early earth and heated them, producing a coenzyme called pantetheine, which is a precursor to a biochemical that helps metabolize nutrients. In another experiment, he exposed methane and nitrogen gas to a spark discharge, producing a compound that reacted with water to form cyanoacetaldehyde. Placing this last compound in a concentrated solution of urea—mimicking a drying beach or a lagoon evaporating from the edges—yielded DNA and RNA bases.

OTHER PREBIOTIC SIMULATIONS

The Miller experiment paved the way for others to attempt to model various conditions under which life might have originated.

Energy Reactions on a Rooftop Tom Waddell, professor of chemistry at the University of Tennessee in Chattanooga, decided, like Miller, to "set up a simple experiment, watch what happens, and let nature teach us." His focus was the energy reactions of metabolism.

Waddell recreated steps of the Krebs cycle, the complex circle of chemical reactions at the center of the pathways that extract energy from nutrient molecules. His hypothesis: Solar energy drove the precursors of some of these reactions on the early earth. Waddell's setup was rather low-tech—he placed chemical intermediates on a sunny rooftop, waited, and observed. In one experiment, a chemical near the "end" of the cycle, oxaloacetic acid, broke down to release citric acid in the presence of sunlight. Citric acid is precisely the chemical that begins the cycle.

Being able to replicate part of the biochemistry of life on a rooftop, or finding it in a meteorite, isn't proof of anything, just as Miller didn't create life. But it suggests ways that biochemistry could have begun and changed. "The Krebs cycle might have originated as an aqueous photochemical reaction on the primitive earth, both before and during the evolution of the first cells. These photochemical reactions might have been the ancestors of some of the steps of the modern Krebs cycle, that were 'recruited' by evolving cells," says Waddell. Gradually, as cells grew more complex, protein-based enzymes were able to substitute for the sun-inspired reactions. "If the hypothesis is correct, we should be able to find microorganisms that still do these pre-Krebs photochemical reactions," he adds.

Mimicking a Membrane A living cell must be separated from the environment because the proportions and organization of its chemicals differ from those of rock or seawater. Biological membranes provide this partitioning, forming two-way selective gateways that retain vital materials, eject wastes, and receive signals that trigger activities inside the cell. These membranes are complex in modern organisms, but simpler versions may have formed spontaneously from precursors on the early earth, given energy input.

A biological membrane is composed of phospholipids. An individual phospholipid molecule is a globular structure with two tails. The globular end includes organic groups and phosphate (phosphorus and oxygen atoms), and the two extensions are fatty acids. What enables these lollipop-shaped chemicals to form barriers is that the globular part is attracted to water, and the tails are repelled by it. When immersed in water, phospholipid molecules aggregate and contort in a way that exposes their *hydrophilic* ("water-loving") parts to water while shielding their *hydrophobic* ("water-hating") parts. The result: a bilayer.

Sound waves can be used to force phospholipids to roll up into bilayered spheres, called *liposomes,* which can serve as a simple model of a cell membrane. But a real cell membrane is enormously complex, with thousands of proteins embedded in, displayed on, or extending through the bilayer. Some of the proteins are attached to sugars or lipids (fats). Researchers can include some of these additions in their liposome recipes, mimicking the specialized functions of cell membranes. But it's a daunting task.

Materials scientist Deborah Charych and biologist Jon Nagy, at the Lawrence Berkeley National Laboratory, custom design liposome-based "artificial cell membranes" to study one function at a time. Their liposomes, which are blue, turn red when sugars and lipids on their surfaces bind viruses. The researchers, who call their invention "molecular Velcro," based it on the observation that viruses often bind to sugars. Their liposomes can serve as diagnostic tests for specific viruses—as well as very simple models of cell membranes.

A Seafloor in the Lab With recent discoveries of diverse life around deep-sea hydrothermal vents, prebiotic simulations have shifted from "primordial soups" to recreations of other conditions. John Holloway, a chemist at Arizona State University in Tempe, has devised a laboratory version of a black smoker. It consists of two linked glass cylinders under high temperature and pressure. The first cylinder contains crushed lava and seawater. The lava contributes iron, sulfur, and silica, and the seawater has dissolved carbon dioxide. The water is heated to 371° Celsius (700° Fahrenheit), but cannot form vapor because it is under high pressure. This superheated water leaches minerals from the lava. A titanium tube connects the first cylinder to the second, which contains artificial seawater. The contained chemicals react, producing a variety of simple organic molecules. Jay Brandes has also built a model of a deep-sea hydrothermal vent, in which he exposed nitrogen compounds and water to an iron-containing mineral. Under high temperature and pressure, the iron catalyzed reactions that produced ammonia (NH_3), one of the key ingredients in Miller's simulated slice of a hypothetical ancient world.

AFTER THE BUILDING BLOCKS FORM: POLYMERIZATION

Prebiotic simulations and chemical analyses of the Murchison meteorite have shown that a variety of simple organic compounds can form under conditions that were likely to have been present on the early earth—and some of these compounds are those found in organisms, such as amino acids, sugars, and nucleotides. Forming such building blocks is the first stage in the transition of chemistry to biology; linking, or *polymerizing,* those building blocks is the second stage.

How did single amino acids or RNA and DNA pieces join amidst the turbulence of the early earth? Minerals may have played a prominent role, serving as both a scaffold and a template, which is a mold that aligns building blocks into sequences. Biochemists Leslie Orgel of the Salk Institute for Biological Studies and James Ferris of Rensselaer Polytechnic Institute create models of primordial polymerization on a common type of clay called *montmorillonite,* which consists of alternating layers of the minerals alumina and silica (Figure 2.4).

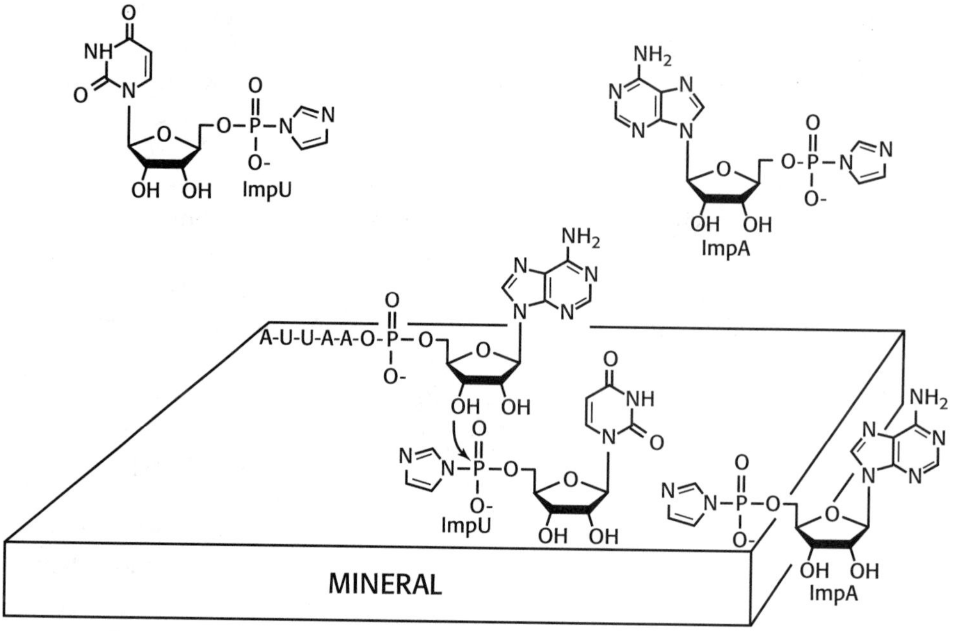

FIGURE 2.4

The building blocks of some of the chemicals of life may have been linked together on clays such as montmorillonite.
Courtesy of James Ferris.

Clay is an inviting site for the aggregation of future biomolecules for several reasons. First, it provides a greater surface area than other minerals. The positive charges on this surface attract and hold the negatively charged units of nucleic acids. Second, clay offers a relatively protected, dry surface that can counter the tendency of water to break apart the forming polymers. Third, not only does the clay hold RNA bases of a single type in place at an angle that eases formation of the phosphate bonds that link them, but it also can foster formation of a complementary strand. (Adenine binds with uracil, and guanine with cytosine.) A molecule of five adenines anchored to clay might have attracted five uracils, which then linked together, forming a short double nucleotide chain. Such an action may have been crucial in the evolution of DNA from RNA, an event we will return to soon. So far, researchers have used clay to coax the formation of 55 units, or "mers," of a single RNA base type, and their complements, in the lab. (Numbers are used to indicate specific-sized polymers, such as a "12-mer" to designate a chain of a dozen single units, or monomers.) Looking at a more complex situation, Ferris imagines "a community of interacting oligomers on mineral surfaces" forming from ingredients readily available in the environment.

Surface clays might not have been the only sites of early polymerization of biomolecules. Pyrite (iron sulfide, also known as "fool's gold") is another mineral that might have been important in the evolution of biochemistry. In 1988, Günter Wächtershäuser, an organic chemist and patent attorney practicing in Munich, suggested that life began on mineral surfaces near deep-sea hydrothermal vents. Here, iron and nickel sulfide minerals from underwater volcanoes may have mixed with carbon monoxide and other gases to yield organic molecules. Prebiotic simulations support the idea.

To a glass-enclosed apparatus, Wächtershäuser added amino acids to hydrogen sulfide and carbon monoxide gases, plus iron and nickel sulfides. He knew that when iron sulfide forms, it releases energy that can be used to assemble an amino acid and to activate it so that it can bond to another amino acid. Conditions in Wächtershäuser's simulation were like those in a hydrothermal vent: water present, oxygen absent, temperature high, and a pH of 7 to 10 (basic)—what he terms "geochemically relevant conditions." After 1 to 4 days in the apparatus, the amino acids linked into groups of twos (dipeptides) and, less commonly, threes (tripeptides). Because the water breaks down the dipeptides rapidly, they must form quickly and easily in order to accumulate as they do. Although it is a long way from short stretches of the same amino acid to the long and complex proteins of life, Wächtershäuser's scenario suggests a way that proteins could have formed on the early earth.

PRELUDE TO AN RNA WORLD

Thanks to simulation experiments, we understand how life's building blocks could have formed on the early earth and linked into short chains. But how did these macromolecules become a protected system, capable of replicating and of changing?

WHY RNA?

Today, the focus of investigation into how chemicals became life is on RNA and similar molecules. Many biologists think that in a time and place before cells, in an "RNA world," this nucleic acid or something like it reigned because it was the only molecule that could both carry information and reproduce. Eventually, a living world based more on DNA and proteins replaced the RNA world. Harvard University's Walter Gilbert coined the term "RNA world" in a one-page paper in *Nature* in 1986, although the idea first arose a quarter century earlier.

RNA wasn't always at center stage in the discussion of how life arose. For many years, thoughts of life's chemical origins focused on proteins, largely because we knew so much about them. A chicken-and-egg dilemma arose. Proteins are synthesized following DNA instructions, but DNA cannot be replicated or accessed without various guiding enzymes and other proteins. The paradox was that DNA and protein synthesis depend on each other—how could one have appeared before the other?

Inklings of a starring role for RNA began shortly after the genetic code was deciphered, in the early 1960s. The genetic code is the correspondence between information in a DNA nucleotide base sequence and information in a protein's amino acid sequence. Several types of RNA carry out protein synthesis. The base sequence of a messenger RNA (mRNA) molecule bears the information encoded in a gene. The mRNA attaches to a globular structure called a ribosome, itself built of many proteins and RNA molecules called, fittingly, ribosomal RNA (rRNA). The very crux of the translation from nucleic acid into protein lays in yet a third type of RNA, called transfer RNA (tRNA). It is a connector, a keylike molecule that grabs a particular amino acid on one end as its other end aligns with a specific triplet of bases in the mRNA (see Figure 3.2 in Chapter 3). Transfer RNAs (tRNAs) bring their cargo to mRNA strands, and the aligned amino acids link. The protein is then released and folds into the characteristic shape vital to its function. Protein synthesis is both fast and efficient. As a newly synthesized protein is peeling off of the mRNA, that mRNA is already being translated again, and the released tRNAs recycled.

RNA does more than assist in synthesizing protein, and its multiple roles suggest that it is, or is simlar to, the molecule or molecules that began life. RNA nucleotides are essential cofactors that help the enzymes in energy reactions, and a bit of RNA is required to start DNA replication. Because it is a single strand, RNA can fold into all sorts of shapes, making all sorts of functions possible. Plus, chemically speaking, a DNA building block can be synthesized from RNA, but not the other way around, making DNA what one researcher calls "a historical afterthought" compared with the more important RNA.

About all that seemed to be missing from RNA's skills was the ability to catalyze certain chemical reactions—such as its own replication. If RNA could act like an enzyme, it might have been able to start life. "In the late 1960s, Carl Woese [see Chapter 3], Francis Crick, and Leslie Orgel, independently and within a year of each other, said 'maybe there is a solution to the chicken and egg problem, and that is that RNA served both functions'" recalls Gerald Joyce, of the Scripps Research Institute. But the idea of an RNA world—what Joyce calls "an organizing principle" rather than an actual place or time—awaited more evidence of what RNA could do. More than a decade later, experiments revealed RNA's catalytic talents.

In the early 1980s, Thomas Cech at the University of Colorado and Sidney Altman at Yale University, in work that would earn them the 1989 Nobel Prize in chemistry, showed that RNA does indeed have enzyme activity, catalyzing reactions that cut other RNAs. Other catalytic capabilities were subsequently discovered. This recognition of RNA enzymes—ribozymes—intensified interest in the RNA world and stimulated development of a technology based on nucleic acid sequences that "evolve" in a test tube.

EVOLUTION IN A TEST TUBE

Naturally occurring ribozymes are sometimes called "molecular fossils" because they may be remnants of a time when RNA-based enzymes orchestrated biochemistry. The ribozymes discovered so far either make or break bonds between phosphorus and oxygen atoms in other RNA molecules or polymerize amino acids. But can RNA catalyze other types of chemical reactions vital to life?

An experimental approach called *in vitro selection and evolution* tests the limits of what RNA molecules can do. Invented independently by three groups of researchers in 1989, the technique entails making huge numbers of different RNAs, selecting out those that demonstrate a particular activity, and then screening those for a more specific activity. This refining process is repeated over and over to select a very specific type of molecule (Figure 2.5).

One of the first *in vitro* selection and evolution experiments, by Jack Szostak at Massachusetts General Hospital and Charles Wilson of the University of California at Santa Cruz, used a DNA synthesizer to make 500 trillion RNA molecules. Each RNA was 112 bases long, and the molecules differed from each other in the sequence of the 72 bases in the middle. All the RNAs were exposed to the vitamin biotin, and a few stuck to it. The DNA synthesizer then made many copies of the bound RNAs, but changed their sequences slightly. This second generation of RNA molecules was bound to biotin again, and those that held most strongly were selected again. The researchers continued through several cycles of selection until they had a few RNAs that bound incredibly quickly and tenaciously. Such an attraction was so strong that it could only be possible if the RNA catalyzed the binding. In this way, Szostak and Wilson discovered RNA molecules that could catalyze formation of a bond between a carbon and a nitrogen to attach themselves

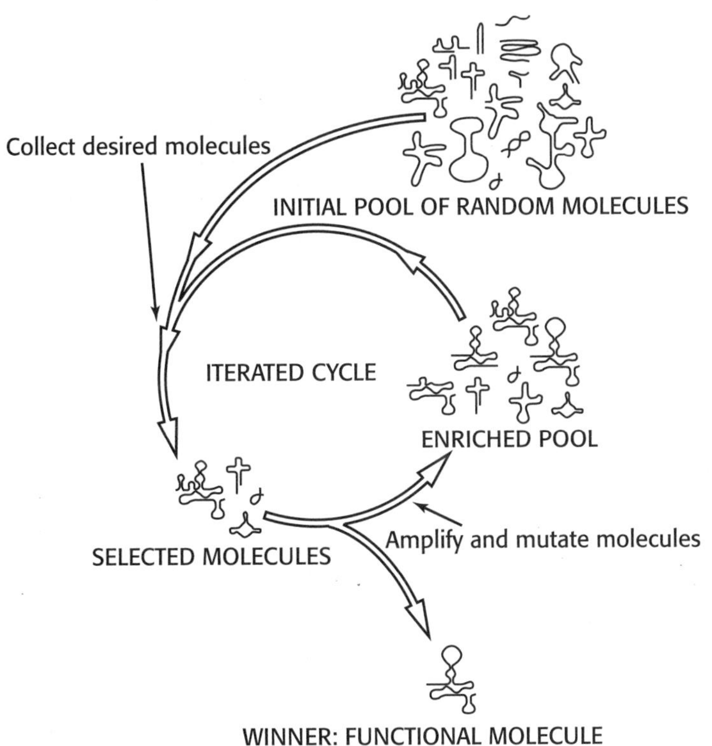

FIGURE 2.5

RNA molecules "evolve" in a test tube when those with a particular activity are selected in repeated cycles of change. In origin-of-life research, such experiments reveal what RNA can do. In a more practical application, as a technology, directed molecular selection could be used to develop new drugs with highly refined mechanisms of action.

Reproduced with permission from Landweber LF, et al. Ribozyme engineering and early evolution. BioScience *1998;48:97. Copyright 1998 American Institute of Biological Sciences.*

to biotin. This is an important ability. Long ago, an RNA that could have catalytically linked itself to something in the environment could have created a new type of molecular assembly.

The power of *in vitro* selection and evolution continues to grow. In another experiment, Peter Unrau and David Bartel of the Whitehead Institute took an almost-unimaginable 1.5 quadrillion lab-made RNAs and gradually homed in on those that could attach a sugar already bound to a phosphate to a molecule very similar to the RNA base uracil.

In a more general sense, these types of experiments are revealing that RNA can do much more than we thought it could. Dozens of different ribozymes have been created in this manner, many more than we know to occur naturally. Some ribozymes can even polymerize RNA nucleotides, an activity that certainly would have been crucial in an RNA world. However, *in vitro* selection and evolution experiments do not necessarily recreate what actually happened long ago. Natural evolution is not orchestrated by a creator (or experimenter), but instead reflects differential survival of certain naturally occurring variants under particular environmental conditions.

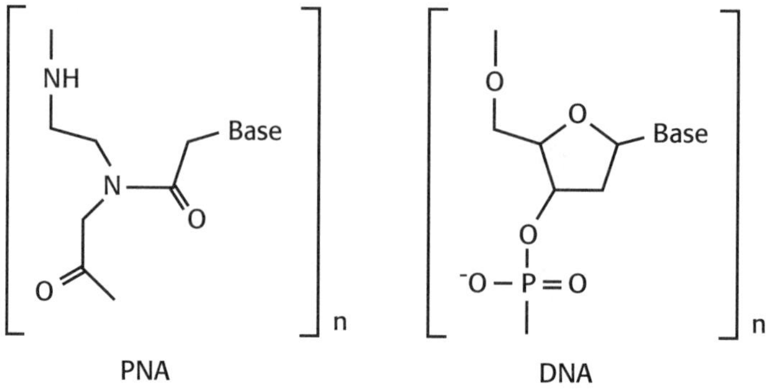

FIGURE 2.6

A peptide nucleic acid has an amino acid as part of its backbone. Because it can bind to single strands of RNA, DNA, and other PNAs, it may resemble a molecule that participated in the start of life.

A PRE-RNA WORLD?

Some researchers are so excited about the idea of an RNA world that they optimistically call its inhabitants "ribo-organisms." Yet as versatile as RNA is, it may not have been the immediate precursor to life. Explains Stanley Miller, "I don't think the first genetic material was RNA. RNA has its problems—the sugars are unstable, synthesizing the nucleosides [bases with attached sugars] is difficult, and although there are attempts at self replication that work better than we may have hoped, it is not good enough. Something is missing."

If RNA was too difficult to make, and too unstable if it did form, perhaps another type of polymer started life. "There must have been a first informational system on the primordial earth. The world is divided into those who say it was RNA, and those who say no, because it is hard to make nucleotides, and the conditions on the prebiotic earth were not favorable for that. If there had to be something before RNA, was there just one something, or several? One can think of one genetic system inventing RNA, or another inventing another inventing another, and so on. I would bet there were not too many intermediates, but I'd be surprised if there was only one," says Leslie Orgel.

One candidate for the basis of a pre-RNA world comes from biotechnology. Antisense molecules are synthetic nucleic acids with sequences complementary to those of known RNA sequences. They are being developed to silence the expression of genes that cause inherited or infectious disease. From such antisense research in 1991 arose a synthetic chemical called a peptide nucleic acid, or PNA (Figure 2.6). Like a nucleic acid, it has four nucleotide bases, but its backbone contains a simple amino acid, glycine (a component of Stanley Miller's "soup."). A single strand of PNA strongly binds to a single strand of RNA, DNA, or another PNA, and it is stable if one of its ends is blocked. Thus, in a reversal of the usual direction of research from scientific principle to technology, a product of technology has added a new dimension to origin-of-life investigation.

Of course we can't know if PNA itself was on the early earth or, if it was, precisely what it did. But its existence, even if only in the lab, suggests that we should look beyond the familiar DNA and RNA in considering the chemicals that led to life.

BETWEEN THE RNA WORLD AND THE FIRST ORGANISM

We don't know what transpired between the time that nucleic acid and protein precursors formed and began interacting to build metabolic pathways and self-replicating systems. But we can imagine what might have happened (Color Plate 1).

Over time, nucleic acid pieces would have continued to form and accumulate, lengthening and increasing in sequence complexity because of the molecule's capacity to change. When a single strand of RNA or DNA pulls in complementary bases to build a new strand, mistakes occur, changing the sequence the next time the nucleic acid replicates. If the change is helpful in the particular environment, it persists and is perpetuated. Some of these early changes may have enabled the gestating biochemical systems to manufacture needed chemicals such as nutrients, formerly obtained directly from the environment. Metabolism would thus have arisen as a consequence of mutation and natural selection. But other forces may have been at work too.

Once a community of nucleic acids existed, new combinations of sequences might have come not from the vertical transfer from one generation of molecules to another, but from lateral transfer—the movement of bases from one nucleic acid molecule to another. The genomes of today show ample evidence for such mobility: Microorganisms harbor DNA sequences of other microbes not at all similar to them. Our own chromosomes bear DNA sequences from viruses that must have nestled there in our distant ancestors. And within cells of many species, some genes literally move from one place to another. Certain plant genomes are riddled with such "jumping genes," and similar nucleic acids might have been vital in life's beginnings. Picture how it might have happened.

Long ago, as PNA or RNA strands collected under protective lipid linings and tapped energy from underlying minerals, lateral gene transfer probably occurred. One researcher suggests that it was something akin to one collection "eating" another. Carl Woese, the researcher from the University of Illinois who is profiled in the next chapter, calls these early assemblages "progenotes." He envisions a time of genetic promiscuity of a sort, with feverish lateral gene transfer and heightened mutation rates generating an entire community of different progenotes. Then, at some point, natural selection intervened, favoring persistence of certain progenotes and dampening the fluidity that continually created new ones. Woese likens the scenario to a heated mixture that is cooled, allowing certain substances to precipitate. Similarly, as the rampant mutation and gene transfer slowed, certain cell subsystems "crystallized out." Somehow, from this first almost-living community, came the form, or forms, that would eventually give rise to the first cells. In Box 2.1, several key players in origin-of-life research voice their views on how this transition from chemistry to cell might have happened.

A FINAL WORD ON CONTROVERSY

Origin-of-life researchers are the first to admit that they don't have all the answers—or any, for that matter. They do so in their publications, in their lectures, and to the media. Stanley Miller carefully explained his attempt at synthesizing possible prebiotic compounds, yet read to his utter amazement in the newspaper how he'd tried to create life. *In vitro* evolution is discredited as being too manipulated to offer any clues to how life began. Many unedited and unreviewed articles on the Web list objections to origin-of-life

BOX 2.1

HOW DID IT HAPPEN?

ORIGIN-OF-LIFE RESEARCHERS SPECULATE ABOUT HOW THE FIRST CELLS AROSE

"The origin of life is the origin of evolution, which requires replication, mutation, and selection. Replication is the hard part. Once a genetic material could replicate, life would have just taken off." — Stanley Miller

"I don't like to look at just one stage for the origin of life. Life began as a process which was inanimate, and ended up animate, with many important stages in between. But we don't know what they are." — Carl Woese

"In my model, the first life was a bunch of RNA bound to a mineral surface. It would have to catalyze better than the mineral to make more RNA, then construct phospholipids to form membranes, and other materials, and put them in a bag, isolated from the environment. The first life could have been bound to rock, it wouldn't need a bag. I envision interaction with nucleotides as they floated past, extracting materials from ocean currents. As the system grew more sophisticated, it learned to make more RNA, amino acids, and peptides." — James Ferris

"If you are satisfied initially with an enclosure that concentrates source molecules for biofunctional compounds and catalyzes their buildup, then I think that we already have robust primitive cells, still existing today but biologically unimportant after the 'genetic takeover.' But if you require a more sophisticated combination of metabolism, reproduction, and mutation, we still have a way to go [in constructing a cell]. Such a successful model system would indeed be inspiring for those of us trying to understand the emergence of life, but there will be no guarantee that it has anything to do with reality." — Gustaf Arrhenius

"I imagine an early life form which might only have consisted of a concentration of proteins, perhaps RNA, a few other chemicals such as citric acid, all under a lipid layer attached to a sulfide grain. It might not even have had complete metabolic systems, instead relying upon chemicals generated elsewhere in the vent system to complete some biochemical pathways. This is an organism that would have been quickly supplanted (digested!) by any modern organisms from the same environment." — Jay Brandes

research—such as the chemical constraints on an RNA world, which researchers openly acknowledge—and then conclude that because there are challenges, the view of life arising from simple chemical precursors is invalid. Some critics extend the argument to conclude that since we cannot understand all the steps, life must have been created by a creator. Even worse, if certain school boards and textbook publishers continue to censor mention of origin-of-life research from curricula, many students will not know that these ways of asking questions and exploring nature exist. And that could bring us back to the days of belief in spontaneous generation—life, in all its great complexity, springing from nothingness. It just doesn't make sense.

But origin-of-life research does not attempt to provide answers, to definitively demonstrate how it was. Instead, it seeks to discover how it might have been, given what we know today about how nature works. Without a time machine, we cannot know whether we have reconstructed what really happened. "You can put a variety of components together, and maybe it lives and reproduces, but it may not be a plausible prebiotic process. You can take all the enzymes and DNA you need, put them in a test tube, and something forms. Maybe you've recreated life, but it is not the same as the historical process because you used purified compounds," says Miller.

Jay Brandes, like many other origin-of-life researchers, sums up the sometimes frustrating limitations of the discipline. "The only truth in the field is that no one knows how life began. There have always been questions that we don't know the answers to, but by hard work and diligence, we have answered a great number of them over the history of mankind. I think that the origin-of-life question is one that we have to try to understand if we are ever to make sense of our place in the universe. And along the way, we should find out some interesting things about ourselves!"

Clues from geology, paleontology, prebiotic simulations, and *in vitro* evolution do not prove that life originated from complex collections of chemicals, because scientific inquiry and investigation cannot prove anything. But the evidence for chemical evolution is certainly more compelling, in a scientific sense, than that for a special creation, spontaneous generation, or vitalism.

REFERENCES

Bada, Jeffrey L., et al. "A search for endogenous amino acids in Martian meteorite ALH84001." *Science* 279:362–365 (1998).

Brandes, Jay A., et al. "Abiotic nitrogen reduction on the early earth." *Nature* 395:365–366 (1998).

Ferris, James P. "Life at the margins." *Nature* 373:659 (1995).

Huber, Claudia, and Gunter Wachtershauser. "Peptides by activation of amino acids with CO on (Ni,Fe)S surfaces: Implications for the origin of life." *Science* 281:670–675 (1998).

Joyce, Gerald F., and Leslie E. Orgel. "The origins of life—a status report." *American Biology Teacher* 60:10–12 (1998).

Landweber, Laura, et al. "Ribozyme engineering and early evolution." *BioScience* 48:94–102 (1998).

Lazcano, Antonio, and Stanley L. Miller. "The origin and early evolution of life: Prebiotic chemistry, the pre-RNA world, and time." *Cell* 85:793–798 (1996).

Lewis, Ricki. "Scientists debate RNA's role at beginning of life on earth." *The Scientist,* 31 March 1997, 11.

Lorsch, Jon R., and Jack W. Szostak. "*In vitro* evolution of new ribozymes with polynucleotide kinase activity." *Nature* 371:31–36 (1994).

Miller, Stanley L. "Production of amino acids under possible primitive earth conditions." *Science* 117:528–529 (1953).

Miller, Stanley L. "Peptide nucleic acids and prebiotic chemistry." *Nature Structural Biology* 4:167–168 (1997).

Miller, Stanley L. "Peptide nucleic acids rather than RNA may have been the first genetic molecule." *Proceedings of the National Academy of Sciences* 97:3868–3871 (2000).

Mojzsis, S. J., et al. "Evidence for life on Earth before 3,800 million years ago." *Nature* 384:55–59 (1996).

Orgel, Leslie E. "The origin of life on Earth." *Scientific American,* October 1994, 77–83.

Pace, Norman. "Origin of life—facing up to the physical setting." *Cell* 65:531–533 (1991).

Robertson, Michael P., and Andrew D. Ellington. "How to make a nucleotide." *Nature* 395:223–224 (1998).

Shopf, J. William. "The cradle of life." Princeton University Press (2000).

White, Owen, et al. "Genome sequence of the radioresistant bacterium *Deinococcus radiodurans R1.*" *Science* 286:1571–1577 (1999).

Woese, Carl. "The universal ancestor." *Proceedings of the National Academy of Sciences* 95:6854–6859 (1998).

GOING OUT ON A LIMB FOR THE TREE OF LIFE

Science is based on evidence. Sometimes, though, evidence can be so contrary to accepted thinking that its discoverer is not taken seriously or is even ridiculed. So it was for Carl Woese, a University of Illinois physicist turned microbiologist who, by looking at cells in a different way, deduced and demonstrated that the units of life come in three basic varieties, not two. In addition to shaking up centuries-old ways of classifying organisms, Woese's experience is perhaps the clearest example ever of the frustration of trying to convince ever-skeptical scientists that a new perspective can sometimes reveal truth.

IN THE LEGACY OF GALILEO

Ideas that once seemed preposterous often come to form the backbone of a science. Consider the plight of Galileo (1564–1642), a man of many talents who today would be considered primarily an astronomer. Looking through a telescope that he invented, Galileo realized that the accepted view of Earth as the center of the solar system couldn't possibly be correct. If this was so, then why did Venus appear in phases, as did the moon? How could Jupiter have its own moons, yet revolve around Earth? His observations, Galileo felt, supported the idea that the planets orbit the sun, which Polish astronomer Copernicus had proposed in 1512. Copernican theory countered the prevailing view of second-century A.D. Greek astronomer Ptolemy that Earth was the hub of celestial activity.

Galileo published his thoughts in a letter in 1613 and in a book shortly thereafter, and spoke publicly about his ideas. In 1616, the Roman Catholic Church convened a panel of "inquisitors," whose charge was to identify and punish heretics. The panel was to consider the ideas and behavior of Galileo. The inquisition didn't brand him a heretic (yet), but forbade him to "hold or defend" Copernican theory. It wouldn't be the first time, nor the last, that scientific fact would be denied because it didn't fit with existing beliefs.

But Galileo would not stay silent. In 1633 he published a treatise entitled "Dialogue on the Two Chief World Systems" that set up an imaginary debate between a Ptolemy supporter whom he called "Simplicio" and a more intelligent follower of Copernicus. The Pope, thinking Simplicio was meant to be he, hauled Galileo before yet another inquisition. This time the outspoken astronomer was found guilty of disobeying the previous order to cease saying that the sun was the center of the solar system. Galileo was permitted to serve his punishment of life imprisonment at a villa due to his age and poor health. Eventually, evidence exonerated Galileo.

In 1979, Pope John Paul II called for a reexamination of the "Galileo case," in light of the long-ago acceptance of the fact that the planets orbit the sun, and to provide guidelines for responding to future scientific discoveries that seem to contradict information in the Bible. A scholarly commission was convened in late 1981 and did not issue a report until 1992, so complex were the issues represented by Galileo's specific case. In addition to exonerating the man, the commission declared that faith and science are two different ways of knowing. Said the pope in 1992:

> The Bible does not concern itself with the details of the physical world, the understanding of which is the competence of human experience and reasoning. There exist

> two realms of knowledge, one which has its source in Revelation and one which reason can discover by its own power. To the latter belong especially the experimental sciences and philosophy. The distinction between the two realms of knowledge ought not to be understood as opposition. The two realms are not altogether foreign to each other, they have points of contact. The methodologies proper to each make it possible to bring out different aspects of reality.

Science inches forward by discoveries that challenge what we thought we knew (see Box 3.1). This is what Carl Woese did when he described and grouped microorganisms according to characteristics that classifiers of life hadn't considered before, and in so doing identified an entirely new type of organism.

Often new tools make possible the discoveries that are, in hindsight, termed *paradigm shifts* because they change thinking in a major way. Galileo had his telescopes, Leewenhoek his microscopes; Woese compared gene sequences among microbes. We look first at how he wed molecular genetics and evolution to shed light on the common ancestors of all life today, and then consider the implications of his discovery.

INSPIRATION FROM THE GENETIC CODE

Carl Woese's entry into biology was gradual (Figure 3.1). He majored in physics, spent a short time as a medical student, and settled on biophysics in graduate school. He spent five years doing postdoctoral research at Yale University before getting his first "real" job at General Electric's Knolls Research Laboratory in Schenectady, New York. The time was the early 1960s. While the civil rights movement, the Vietnam conflict, and the British invasion of music dominated headlines, biological science was undergoing a quiet revolution of its own.

In 1953, James Watson and Francis Crick, building on the experimental work of others, deduced the structure of DNA. The molecule of heredity is a double helix built of nucleotide chains that are held together by the specific pairings of four types of nitrogen-containing bases: adenine (A) with thymine (T), and guanine (G) with cytosine (C). Their seminal 1953 paper in *Nature* ended with the tantalizing comment, "It has not escaped our notice that the specific pairing we have postulated immediately suggests a possible copying mechanism for the genetic material." That is, DNA's twisted-zipper structure parts and pulls in new building blocks to create two double helices from one. Elegant experiments verified Watson and Crick's model for how DNA replicates.

But knowing DNA's structure and how it replicates left unanswered an important question: How do the 4 types of DNA bases encode the 20 types of amino acids in life, instructing the cell to build proteins? Woese and many others became sidetracked from physics, chemistry, and mathematics to focus on this compelling question, brainstorming candidate codes and experimentally testing them. The sense that they were exploring a very fundamental aspect of life gave the pursuit of the genetic code an aura of science fiction, Woese recalls.

Intense research revealed the players of protein synthesis in those early years of the 1960s. A long messenger RNA (mRNA) molecule bears the information encoded in a DNA strand; a globular ribosome built of proteins and ribosomal RNA (rRNA) holds the mRNA in place; and the small, key-shaped transfer RNA (tRNA) connects an mRNA with the amino acid it specifies. When all the pieces come together in the right way, amino acids are brought in one at a time on tRNAs and lined up opposite the complementary sequence of mRNA, linking as they arrive to form a growing chain that will eventually be long enough to be

BOX 3.1

THINGS WE THOUGHT WE KNEW

Sometimes we are fairly certain of a scientific "fact," until further work—thanks to a new tool, new knowledge, or a different perspective—shows otherwise. Science seeks truth, but that truth is always tentative. Here are some examples, in no particular order, of scientific facts about life that were altered or amended as we learned more.

THE WAY IT WAS ...	THE WAY IT IS (FOR NOW...)
Life can arise from nonliving matter.	*Life comes from preexisting life.*
Human sperm contain a tiny human.	*Human sperm contain a nucleus, mitochondria, and not much else.*
Living in crowded cities causes tuberculosis.	*Bacterial infection causes tuberculosis, but crowded living conditions foster its spread.*
Smoking does not harm health.	*Smoking causes lung cancer in some people.*
Stress causes ulcers.	*A bacterial infection, perhaps in addition to stress, causes ulcers.*
Protein is the genetic material.	*DNA is the genetic material.*
Humans have 48 chromosomes.	*Humans have 46 chromosomes.*
Viruses cause spongiform encephalopathies such as kuru, scrapie, and "mad cow disease."	*Proteinaceous infectious particles (prions) cause these disorders.*
Amyl nitrate ("poppers") causes AIDS.	*A virus causes AIDS.*
Men and women breathe differently. Men use their diaphragms, whereas women breathe from their top ribs.	*Men and women breathe the same way The tight corsets of the nineteenth century restricted women's breathing.*
Genes are stationary.	*Some genes can move among the chromosomes.*
Radium is safe.	*Women who used radium-laced paint to make watchfaces developed cancers.*
Genes are continuous sequences of amino acid–encoding bases.	*Genes contain segments (introns) that are not translated into protein.*
RNA serves only to translate a gene's sequence into protein.	*RNA has a variety of functions.*
A cell's fate is determined by which of the three layers in the embryo it derives from.	*Certain cells can change how they specialize.*
All enzymes are proteins.	*RNA can function as an enzyme.*
Fever is a sign of illness.	*Fever is part of the immune response.*
A mammal can never be cloned.	*Hello Dolly!*

FIGURE 3.1

Carl Woese.
Photo courtesy of University of Illinois News Bureau.

considered a protein—the gene's product (Color Plate 2). Experiments revealed that the genetic code is triplet, with three mRNA building blocks encoding one amino acid. But the big challenge was to decipher which mRNA triplets specify which amino acids.

In 1961, Marshall Nirenberg, a biochemist at the National Institutes of Health, devised a way to simulate the gene-to-protein process in a test tube, using ingredients from shattered bacteria and synthetic RNA molecules consisting of a single base type, such as UUUUUU and CCCCCC. (RNA contains uracil [U] instead of thymine.) By seeing which amino acids the four such simple RNAs linked, Nirenberg and his coworkers discovered the first four pieces of the genetic code. By using RNAs consisting of two different bases (such as UCUCUCUC) and then three (such as CAGCAGCAGCAG), they and other investigators eventually matched 61 of the 64 possible RNA triplets to one of the 20 amino acids of life, the remaining three codons signifying "stop." The code was broken. The key researchers had a group called "The RNA Tie Club," begun in the 1950s. They would induct a new member each time someone discovered another piece of the genetic code puzzle. (Today, many nonscientific publications confuse genome projects with "breaking genetic codes." The genetic code, which is universal to all species, was broken in the early 1960s.)

Woese worked on a version of a genetic code based more on the chemical interactions between RNA and amino acids than the cryptographic code that Crick and Nirenberg and others proposed. His published ideas attracted attention. He gave talks and made important contacts at meetings, as enthusiastic young scientists do, and in 1963 the University of Illinois coaxed him away from GE. In that era following the success of the Russian satellite *Sputnik* and the ensuing romance of the early U.S. space program, money poured into scientific research, and Woese had his choice of funding from the National Institute of Health and the National Science Foundation. "Think now of what an assistant professor has to go through in securing grants! It was a really fortunate time to be in science," he recalls.

ENTER EVOLUTION

In parallel to the emerging details of the genetic code, the field of molecular evolution was also gestating. The basic idea was that evolutionary relationships could be inferred between pairs of modern species by comparing the amino acid sequences of their proteins or the base sequences of their DNA or RNA molecules. This conceptual approach assumes that the most logical explanation for sequence similarities between pairs of species is descent from a common ancestor. Deciphering gene sequences thus became not just a way to understand how cells work, but also a tool to answer questions about how life has evolved and diversified, of how species are related to each other.

"A wonderful concept began to emerge, with sequences of globins and cytochromes, and other proteins in different organisms. The proteins were apparently identical in function, yet with different sequences. It was a very novel concept at the time," says Woese. Not only were vital proteins similar among species, but all organisms studied made their proteins the same way, even using the same genetic code.

It was against this backdrop of rapid-fire discoveries that Woese began to think about new ways to compare organisms. His goal was the Holy Grail of biology—to identify the "universal ancestor," an organism that is the most recent to have given rise to all known forms of life. Presumably, finding evidence of this last common ancestor would extend our knowledge further back than what the fossil record reveals, because remains of the earliest cells must have been destroyed long ago. It was a lofty goal that Woese has not yet met, but as often happens in science, he discovered something completely unexpected.

In evolution if something works it tends to persist, Woese thought, a little like an "if it ain't broke, don't fix it" scenario. Therefore, the genes that encode the rRNAs and tRNAs that all organisms use to build proteins should be similar, if not even identical, among species. Great differences in these genes would signify organisms that were not closely related—that is, that had diverged from a common ancestor long ago—because it would take time for the changes to accumulate as the genes naturally mutated. Why not compare the genes encoding rRNA or tRNA, much as protein sequences were being compared, to uncover evolutionary relationships?

Woese remembers when he had the idea. "I dropped everything, and thought about trying to relate organisms to one another through their rRNA sequences. I wrote a letter to St. Francis [DNA codiscoverer Francis Crick], asking 'should I go into evolution?' He wrote back, yes. Francis Crick gave his stamp of approval. I would have done it anyway, but it was nice to have a pat on the back from someone you respect so much when you are starting on a long journey." Thus he set about separating rRNAs from the ribosomes of bacteria. And, with apologies to Jerry Garcia, what a long, strange trip it's been!

THE EXPERIMENTS

In the early 1970s, DNA sequencing had not yet been invented, let alone the automated devices that today decipher strings of many millions of bases a year. Instead, in a technique called autoradiography, researchers used enzymes to cut RNA into pieces (oligonucleotides, or *oligos*), labeled them so that they emitted radiation, and spread the pieces onto a sheet of paper. Then they laid down photographic film over the paper, which would, after several days, reveal the locations and sizes of the oligos. The collection of RNA molecules found in a cell of any species would yield a characteristic pattern of spots on the resulting autoradiogram, providing an organismal bar code of sorts.

Woese knew that the two types of cells—prokaryotic and eukaryotic—produced two very distinctive patterns of rRNA spots. Prokaryotes include bacteria and cyanobacteria (once called blue-green algae); eukaryotes encompass all other species, including multicellular organisms (protista, plants, fungi, and animals). The cells of eukaryotes differ from those of prokaryotes in that the DNA is contained in a nucleus, and many membrane-bound structures called organelles compartmentalize specific cellular activities. Prokaryotic cells are also complex in organization, but are generally smaller and lack nuclei and membrane-bound organelles. If a eukaryotic cell is considered the equivalent of a large department store, then a prokaryotic cell, in contrast, is more like a convenience store.

The lab at the University of Illinois had several tanks in which to run the paper chromatography separations. Woese combed his building, borrowing microbes to test from whomever could spare them. He and a graduate student would then "catalog away," amassing rRNA patterns for yeast (a eukaryote) and various bacteria that were well-known laboratory strains. Then a fellow professor and friend, Ralph Wolfe, suggested Woese look at a bacterium from a more natural habitat, *Methanobacter thermoautotrophica,* or "delta H" for short. This microbe is unusual in that it produces methane, more colorfully known as swamp or marsh gas. "Wolfe's student had gotten it from the local city septic system. Microbiologists commonly go to swamps, pigpens, sewage processing plants and such to find methanogens. Around here we go to Crystal Lake. We wade in, stir it up, and invert a funnel over the stirred up area. Methane gas bubbles up, and we strike a light, and flame comes from the small end of the funnel. That's marsh gas, or methane," Woese recalls. Biologists had been lumping methanogens together because these microbes all can grow in the absence of oxygen by reacting hydrogen with carbon dioxide to produce methane. Yet no one knew how the methanogens were related to each other or to other bacteria.

Delta H would turn out to open the door to the third major group, or domain, of life, known today as Archaea. Woese affectionately calls its members "arkies." He tells with excitement what happened next:

> Along comes our first methanogen, and we made oligos. We had to cut spots out of the paper, redigest them, cut smaller pieces that overlap, and infer RNA sequences. Here I was inferring away on that methanogen, and what the hell, there were two missing spots! That was our first clue that this pattern wasn't normal for prokaryotes. When we'd come to spots that should have been there as part of the prokaryotic signature, there would be no spot, or not the right sequence. The parts of the signature for prokaryotes we looked for were mostly missing.

The group had already obtained such signatures for 40 or so well-known bacteria. When scientists see an experimental result that doesn't fit with what they've seen before, all but the most egotistical generally conclude that they have fouled up in some way. So it was with Woese and his mysterious methanogen.

"A third of the way through, I knew that what I was seeing was either an error, or something I didn't understand. So, as is the way of science, we did the experiments over, and found the same result. The more we looked at the data, the more it appeared that the methanogen was neither a prokaryote nor a eukaryote. We were redefining the methanogen." Soon, they discovered other methanogens that fit the delta H rRNA pattern.

THE ARKIES' DEBUT, IN PRINT

At this point, the saga of the arkies' shaky debut became recorded in the scientific literature as well as in newspaper headlines. In January 1977, Woese reported in *The International Journal of Systematic Bacteriology* the rRNA data for seven common bacteria, all showing the characteristic prokaryotic signature. This paper laid the groundwork. Then, in October 1977, in an article in *The Proceedings of the National Academy of Sciences (PNAS),* Woese and colleagues George Fox, Linda Magrum, Ralph Wolfe, and William Balch described the distinctive rRNA patterns shared among 10 types of methanogens, including delta H. The abstract concluded with the then-outlandish statement that "These organisms appear to be only distantly related to typical bacteria."

The next month's *PNAS* had an article that would precipitate the paradigm shift. In it, Woese and his student George Fox proposed the "three aboriginal lines of descent": the eubacteria (traditional bacteria), the archaebacteria (the methanogens), and the urkaryotes. This last group included eukaryotic cells minus their mitochondria and chloroplasts, which are organelles that were once free-living eubacteria or archaebacteria—another long story. Woese and Fox initially used the word *archaebacteria* because they knew they were going out on a limb and anticipated major objections to subdividing the prokaryotes. The "archae" recognized the fact that the first such organisms identified came from a methane-rich environment, which is similar to early earth conditions, suggesting that these microbes might be quite ancient. The discovery of archaea in deep-sea hydrothermal vents soon supported this idea.

While scientists digested Woese's ideas in *PNAS,* the public read about them on the front and editorial pages of the *New York Times.* Both accounts revealed an astounding misunderstanding of the state of biology circa 1977. Claimed the editorial writer, "Every child learns about things being vegetable or animal—a division as universal as the partition of mammals into male and female. Yet . . . [we now have] a 'third kingdom' of life on earth, organisms that are neither animal nor vegetable, but of another category altogether." The writer had apparently never suffered a bacterial infection, seen an ameba, or eaten a mushroom. *Prokaryote* and *eukaryote,* the terms needed to fathom what Woese was on the verge of suggesting, were not part of the everyday lexicon. The *New York Times'* misguided condescension was foreshadowing.

MAKING THE CASE

In science, alternate hypotheses are set up to challenge unpopular ideas. If the alternate explanations can be disproved, the original hypothesis stands—until it is disproven. To Woese, the most logical explanation for the unique rRNA signatures of the methanogens was that they are as they appear, a group of microbes different from bacteria. But skeptics suggested other explanations. Were the 10 tested methanogens revealing a real difference, or were they merely quirks? More experiments, and more data, were needed to rule out competing hypotheses. Woese and his colleagues soon did just that.

Woese had been working with very small, wall-less bacteria called mycoplasmas for another project. "One was *Thermoplasma,* the oddest of the bunch because it liked acid. I thought it was another mycoplasma," he recalls. But the rRNA analysis said differently. Whatever it was that Woese was seeing, it was more than just a peculiarity of the marsh gas microbes. "*Thermoplasma* fit the methanogen pattern! That was a pleasant surprise," he says. But reaction of the scientific community to the discovery was not so pleasant.

First, Woese hit a publishing roadblock. "Very shortly after *Thermoplasma* happened, we sent it to *Nature,* fast. This was the first nonmethanogen member of the new group! The *Nature* editor said, 'we've published enough of this work,' and turned it down at the first opportunity. I don't think they even sent the paper out for review. It was a 'seen one, seen them all' attitude. That's how popular the idea was," Woese recalls.

The *Nature* rejection was an overt sign of the opposition to the idea of a third form of life. Then subtler signs surfaced. "Most of the objection to the arkies was behind the scenes. In November 1977, when the paper first broke, Ralph Wolfe got a call from a Nobel prize–winning friend, Salvatore Luria, who said, 'Ralph, it is essential that you divorce yourself from this nonsense. It is going to ruin you,'" Woese says.

Luria's hesitancy to accept the hypothesis that some bacteria weren't really bacteria turned out to be fairly typical, and says something about how difficult it is to gain support for a novel idea. "The reason that Luria called was that everybody knew that there were prokaryotes and eukaryotes, and all bacteria were prokaryotes, as if it was taught in church. Salvatore Luria's comments were made because he knew that all bacteria were prokaryotes, and that was it. Well, I knew that too—until that day in June of 1976. At that point, I knew something wasn't right about the prokaryotes, if you believe what you saw. Yet even after we confirmed it, people stuck to the dogma," Woese relates.

As Woese struggled with the semantic problem of just what to call the new group—archaebacteria was proving confusing—he became aware of ridicule. "We weren't taken seriously. People made jokes. At the American Society for Microbiology annual meeting that spring, 1978, we were kidded. Wolfe, a card-carrying member of the society, had to put up with snide remarks. Everyone said archaebacteria were just bacteria."

But Woese and his lab carried on. By the time they'd identified about 50 arkies, they'd switched to a newer, faster RNA sequencing method. The list of organisms continued to grow, and at an accelerated pace. "A lovely group of archaea was emerging," Woese remembers. And it would turn out that they had more in common with each other than just one type of gene sequence.

DETAILS EMERGE

After finding the archaean rRNA signature in *Thermoplasma,* the researchers next identified it among the halophiles, a group of microbes that thrive in salty conditions. The halophiles were known for their unusual lipids, which Woese knew *Thermoplasma* had too. Since the methanogens resembled the halophiles and *Thermoplasma* in their rRNA, Woese wondered if they had the peculiar lipids too. They did! The discovery that all three types of microbes shared key rRNA sequences as well as lipids added evolutionary meaning to the presence of the lipids in *Thermoplasma* and the halophiles. Microbiologists had thought the unusual lipids were in both organisms just by coincidence. The mounting similarities were evolving from aberration or happenstance to telling a story—if investigators would pay attention to the clues.

The characteristics of the archaea have continued to emerge. They are a variable group, with many shapes, reproductive strategies, and ways of acquiring energy. They live not only in hostile (to us) habitats, but in many places, including throughout the layers of oceans, rice paddies, marshes, and in the intestines of large vertebrates.

Exactly two decades after Carl Woese discovered the archaea by looking at pieces of RNA, an avalanche of similar data arrived in the form of the complete genome sequence of *Methanococcus jannaschii,* an archaeon collected by the submarine *Alvin* in 1982 from a "white smoker" chimney bubbling up from the Pacific Ocean. A sample of material from the chimney found its way to Ralph Wolfe, who isolated and characterized the organism and gave some to Woese to catalog its rRNA. The rRNA profile placed *M. jannaschii* with the other methane-producing archaea. In 1996, Carol Bult and her team at The Institute for Genomic Research in Rockville, Maryland, sequenced all 1,664,976 DNA bases that constitute *M. jannaschii's* genetic material.

The organism is a curious mix of prokaryote and eukaryote—and then some. The genes encoding its surface features and most of its metabolic enzymes resemble those of run-of-the-mill bacteria, yet its DNA replication and protein synthetic machinery are more like those of eukaryotes. The lipids in its membranes and cell walls are unique. The biggest surprise was that 56% of its 1738 protein-encoding genes were at the time completely unknown in any prokaryote or eukaryote. On a whole-genome level, *Methanococcus jannaschii* very much appeared to be a third form of life.

Carl Woese was one of many authors of the *Science* paper unveiling *M. jannaschii's* genome, and his picture was prominent in the accompanying news article. Yet he bristles at the idea that, as the news article suggests, the genome work was necessary to confirm what he had shown 20 years earlier with rRNA sequences. "Microbiologists who had been paying attention had an understanding about the archaea long before 1996," he says. The genome work also illustrates the evolution of molecular genetics research from the more inferential and, some would say, creative work of the 1960s and 1970s to today's systematic high-throughput sequence crunching.

Hypotheses ascend to theory status only with the accumulation of evidence and data. The idea of archaea as a third type of cell is clearly becoming a theory. But Woese suggests another factor in the growing acceptance, one reminiscent of Moses wandering in the desert for 40 years to purge his people of those who worshipped idols. "Characters were retiring who'd challenged the idea, and were being replaced with people without the prokaryotic indoctrination. The younger ones have made the difference. Charles Darwin said, 'The road to scientific progress passes through the graveyard.' He also said, 'You don't convert your peers, you educate the next generation.'" But unlike Woese's ideas, those of Darwin met near-immediate acceptance among scientists.

ARKIES AFFECT BIOLOGICAL CLASSIFICATION

The existence of archaea created an instant dilemma for biologists. How did these organisms fit into the accepted five kingdoms of life? One solution was to give them their own, sixth, kingdom. But this forced a contradiction: The differences of the archaea from members of the other kingdoms transcended all the differences between the existing kingdoms. A new, broader taxonomic level was needed, one above kingdoms. Woese and coworkers proposed in a 1990 *PNAS* paper that there are three domains of life: the Bacteria, the Archaea, and the Eukarya (Figure 3.2).

Adding domains atop the taxonomic hierarchy was a major change to the centuries-old field of biological classification. To appreciate the magnitude of invoking domains above kingdoms requires a look at another Carl in another time.

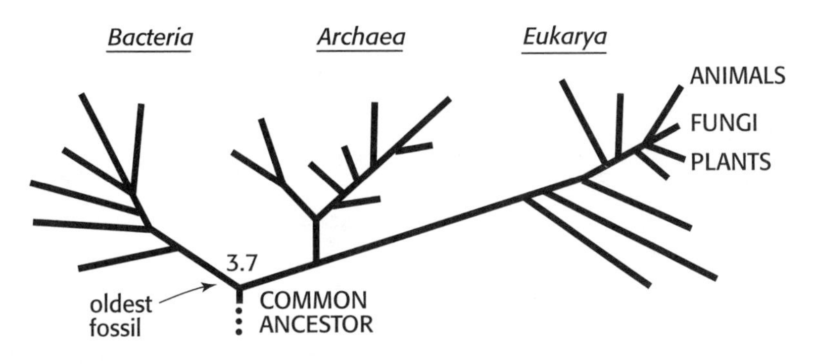

FIGURE 3.2

The three domains of life: Bacteria, Archaea, and Eukarya.

HISTORY OF BIOLOGICAL CLASSIFICATION

Carl Linnaeus (1707–1778) was, by many accounts, a man obsessed. He had an amazing ability to observe, note similarities and differences, and express, analyze, and categorize what he saw in words. In short, he liked to make lists. As a child in Sweden, he and his father nurtured a large garden of exotic plants, which Carl would meticulously arrange into lists and tables. When he was older, he traveled extensively, observed plants, and made more lists, grouping plants by similarities of their sexual parts. In medical school, he learned by composing lists and tables of diseases. Even later, as a professor of botany at the University of Uppsala, he'd make lists of all the botanists he knew, arranging them by a military rank that he assigned, appointing himself general.

Linnaeus started organizing plant life when he was 8 years old. By age 28, in 1735, he published the first edition of *Systema Naturae,* his classification of the "vegetable kingdom." That first edition consisted of 7 large pages. *Systema Naturae* became a lifelong project, and by the tenth edition it had reached 2500 pages.

As the "father of taxonomy," Linnaeus made two basic contributions: Organisms come in distinct types (species), each of which is identified by a unique designation; and each species is also part of a genus, which is a small group of species that share a few characteristics. A type of organism is distinguished by a binomial name, which consists of its genus and species. Linnaeus also assigned higher levels of organization, the class and order. It would be two generations before French anatomist Georges Cuvier (1769–1832) introduced phyla, a designation broader than order (Table 3.1).

Binomial names described types of organisms by what people could see, but did not imply or intend anything about common descent. Linnaeus viewed species as immutable: God had placed two of each kind, a male and female, on the earth, from which their numbers had increased. No new species could arise. Years later, Charles Darwin (1809–1882) disagreed, writing that "our classifications will come to be, as far as they can be so made, genealogies."

Linnaeus wasn't the first to classify life. Even prehistoric cultures probably categorized their nonhuman neighbors in ways that were important to them, such as "safe to eat" or "of medicinal value" or "avoid—danger!" The Greek philosopher Aristotle (384–322 B.C.) was one of the first to systematically catalog life, organizing 500 or so types of animals

TABLE 3.1

Traditional Taxonomic Levels

Level	Human Classification
Kingdom	Animalia
Phylum	Chordata
Class	Mammalia
Order	Primates
Family	Hominidae
Genus	*Homo*
Species	*Homo sapiens*

into hierarchies, albeit subjective ones. He distinguished animals by habitat (land versus water) and by body form, then ranked them, assigning the highest regard to those that were warm and moist. Of course, the most perfect animals had lungs and gave birth to live young, and the most perfect of the perfect was *Homo sapiens.* Although Aristotle recognized the importance of live birthing, he nevertheless placed the male of our species at a higher rank than females, at the pinnacle of all life.

Aristotle's hierarchies sufficed when the types of observable organisms numbered in the hundreds. But in the sixteenth and seventeenth centuries, with the age of exploration taking people to new lands, the lists of new plants, animals, and fungi grew, and a more systematic organizational framework became necessary. Like adding new area codes to generate more phone numbers, biologists needed meaningful umbrella terms under which to place groups of organisms that shared traits that were inherited from common ancestors. Sometimes the value of particular traits in establishing evolutionary relationships was questionable, in light of today's more objective gene and protein sequence comparisons. For example, a listing of plants from the early 1500s justifies a plantain's lofty position in the hierarchy "because more than any other plant, it bears witness to God's omnipotence." Another scheme classified organisms by their "powers" and "temperament."

John Ray (1627–1705) was an English naturalist whose meticulous descriptions and groupings of plants would later help the young Linneaus. In three volumes called *Historia Plantarum Generalis,* published from 1686 to 1704, Ray categorized 18,600 types of plants. He made some interesting blunders, which reflects the fact that early classifications were based on what the classifiers could easily see. Ray grouped together fungi, algae, lichens, and corals, and distinguished animals by their toes, teeth, claws, nails, and hoofs.

Until the invention of microscopes, classifications were limited to plants and animals, with anything of dubious identity, such as the occasional mushroom, considered a plant. But the increasing number of organisms that seemed not quite plantlike, and then the discovery of the world of microbes, bothered biologists. In 1866, German naturalist Ernst Haeckel (1834–1919) suggested a third kingdom, Protista, which would include one-celled organisms that weren't plants or animals. However, this lumped together some very diverse microscopic organisms. So in 1937, French marine biologist Edouard Chatton ventured beyond the kingdom concept to describe prokaryotes and eukaryotes, based on the absence or presence of a cell nucleus. But by 1959, even three kingdoms couldn't contain the

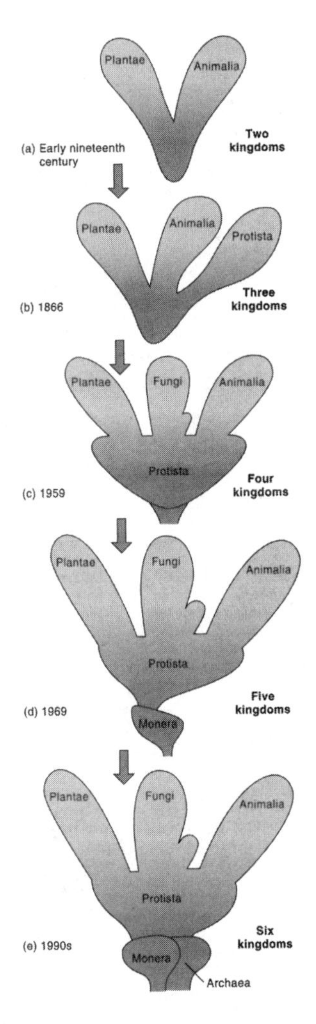

FIGURE 3.3

The evolution of classifying life. At first, biologists classified only what they could see, which appeared to be only plants and animals; fungi were long considered to be plants, and unicellular life was undetectable. A three-kingdom system eventually embraced single-celled life, and a four-kingdom system distinguished prokaryotic from eukaryotic unicellular organisms. The five-kingdom system belatedly recognized the distinctiveness of the fungi. Some biologists used a six-kingdom system during the time that the archaea were first being described, when it began to become apparent that a five-kingdom system would not suffice.
Lewis, R. Life, 3rd ed. Boston: McGraw-Hill, 1998:430. Reproduced with permission of The McGraw-Hill Companies.

burgeoning number of organisms biologists were discovering.

Several four-kingdom schemes prevailed for about a decade, one distinguishing the fungi from the plants, another wedding fungi to Protista and considering bacteria alone. Then in 1969, Cornell University ecologist Robert Whittaker established the five-kingdom system that remained popular until the "arkies" came along. The five kingdoms are Monera (bacteria, which then included the archaea), Protista (single-celled eukaryotes), Fungae, Plantae, and Animalia (Figure 3.3).

DOMAINS ARRIVE—FOR SOME

In 1977, by declaring the methane-producing bacteria distinct from others, Carl Woese necessitated a reevaluation of the reigning five-kingdom system of biological classification. Today the idea of domains has finally taken hold in biology textbooks at the college level, with the lower-level texts still somewhat behind (Table 3.2).

Textbook coverage has mirrored the intellectual evolution of the concept. "Archaeabacteria" first appeared in hesitant footnotes and figure legends, then in the occasional carefully worded sidebar. These mentions typically credited Woese, but presented the idea of a third domain of life in particular, and domains in general, as still somewhat renegade because of the promotion by one individual, something very unusual in science. After a few transitional editions of not knowing exactly where to put the archaea, today's textbooks are literally being rewritten to embrace them as a bona fide third form of life—a third domain, not a sixth kingdom. Perhaps as researchers find more sequence distinctions, a nomenclature based on precise identifying criteria for the archaea will arise, similar to the rules of nomenclature for organic (carbon-containing) chemical compounds. For example, a microorganism with a certain set of rRNA

TABLE 3.2

The Evolution of an Idea

Year	Publisher (Level)	Title and Author	Coverage
1940	McGraw-Hill (college)	*This Living World*, Clark and Hall	Two kingdoms. "[T]he first great divergence of life led to the plant and animal kingdoms. Since that time almost all living creatures on earth have the characteristics of either plants or animals."
1967	W. W. Norton (college)	*Elements of Biological Science*, Keeton	Three kingdoms, but confronts the dilemma of unicellular eukaryotes as different from plants.
1980	Biological Sciences Curriculum Study (high school)	*Biological Science: A Molecular Approach* ("blue version")	In a figure of two-, three-, four-, and five-kingdom schemes, the legend states "A six-kingdom system has also recently been proposed."
1985	Heath (high school)	*Heath Biology*, McLaren and Rotundo	Five kingdoms "is more in keeping with current knowledge than any other classification scheme developed so far."
1989	Holt, Rinehart and Winston (high school)	*Modern Biology*, Towle	Five kingdoms, with box on archaebacteria and Woese.
1993	W. W. Norton (college)	*Biological Science*, Keeton and Gould	Three domains; lists many kingdom schemes.
1994	John Wiley and Sons (college)	*Biology: Exploring Life*, Brum, McKane, Karp	Five kingdoms, but essay says that Woese proposed three domains.
1994	Benjamin/Cummings (college)	*Biology: Concepts and Connections*, Campbell, Mitchell, and Reese	Five kingdoms. "[T]he broadest groups are called the kingdoms of life." Places Archaebacteria with bacteria.
1995	McGraw-Hill (college)	*The Nature of Life*, Postlethwait and Hopson	Three domains, six kingdoms.
1998	McGraw-Hill (college)	*Life*, Lewis	Three domains, five kingdoms.
1998	ITP Wadsworth (college)	*Biology: The Unity and Diversity of Life*, Starr	Six kingdoms; mentions domains briefly.
2000+	Coverage of three domains increases in many texts.		

sequences would be designated an archaeon, much as an organic molecule with a hydroxyl (OH) group is designated an alcohol.

It is the very nature of science to continually question, and not everyone accepts the idea of three domains of life, even now. In an article in *PNAS* published in May 1998, noted biologist Ernst Mayr, professor emeritus at Harvard University, argued against considering the archaea anything but prokaryotes; they are, he insists, just plain old bacteria. But Mayr's self-description as "a student of the eukaryotes" reveals his biases. As evidence that the archaea do not comprise a distinct group, he cites their apparent rarity—biologists know of only 175 or so archaea and 10,000 or so species of bacteria; in contrast, they have identified more than 30 million species of eukaryotes. The numbers just don't make sense, he says. How can 10,175 types of organisms include two basic types of life, while many millions of eukaryotes are all of a kind? According to Mayr's reasoning, since archaea and bacteria are less abundant, they should probably be placed together.

Numbers aren't the only way that Mayr's analysis is subjective. He claims that the visible differences between the two kinds of prokaryotes "is minimal as compared with the differences between, let us say, a bacterium and a plant or animal." And he cites the very failure to recognize the archaea for so long as evidence that they really aren't different from bacteria. Mayr's words, although eloquent, reveal his skewed view, reminiscent of Aristotle's tendency to organize those beasts with which he was personally familiar. Mayr wrote of "the wonderful organic diversity represented by jellyfish, butterflies, dinosaurs, hummingbirds, yeasts, giant kelp, and giant sequoias. To sweep all this under the rug and claim that the difference between the two kinds of bacteria is of the same weight as the difference between the prokaryotes and the extraordinary world of the eukaryotes strikes me as incomprehensible."

In a rebuttal published in *PNAS* two months later, Woese calls these comparisons "eye-of-the-beholder type of diversity." Textbook author William Keeton succinctly pointed out the same subjectivity in his 1967 textbook *Elements of Biological Science:* "Plants are living things studied by people who say they are studying plants (botanists), and animals are living things studied by people who say they are studying animals (zoologists)."

IT'S A SMALL WORLD AFTER ALL

The seemingly greater diversity of eukaryotes compared with other organisms reflects the state of our knowledge, not the reality of nature. Microbiologists readily acknowledge that the pathogenic bacteria that we recognize, which may include some archaea, account for at most 1% of microbial species. Add that to the fact that microscopic life has existed for many millions of years longer than the oaks and orchids, bats and chickens of Mayr's discussion, and the potential for their diversity to be far greater than that of eukaryotes seems obvious.

Prokaryotes' extended time on earth, plus their abundance, suggests that they must be more diverse than eukaryotes, even if we can't see them. They've simply had more time to diverge into new forms. A 1957 study estimated that half of all biomass is microbial; a more recent investigation by microbiologists William Whitman, David Coleman, and William Wiebe at the University of Georgia provides compelling evidence that even that estimate is likely too low. Their work was published in *PNAS* just two months before Mayr's article. The Georgia researchers conducted censuses of many habitats and extrapolated to calculate the number of prokaryotic cells on earth—it's 5 million trillion trillion, or

5,000,000,000,000,000,000,000,000,000,000

The three most heavily populated types of habitats are seawater, the soil, and the interface between soil and sediments. But considering the number of prokaryotic cells that reside on and in eukaryotic organisms is also eye-opening. The human armpit, for example, is home to some million prokaryotes per square centimeter. An average person's skin harbors 300,000,000 microorganisms, and our intestines house many more. A square centimeter of a leaf growing in a rainforest sports 100 billion prokaryotes. A single termite's hindgut contains 2.7 million of them. Multiply that by the 2.4 quintillion termites on earth, and it's easy to see that these "simple" cells far outnumber any visible and familiar beasts or bushes.

With so many unicellular organisms that lack nuclei populating the planet for so long, their natural division into two basic types, as defined by their genes, does not seem all that unlikely.

A FINAL NOTE ON PERSPECTIVE

The discoveries of Carl Woese and their impact on the field of biological classification teach a larger lesson—that the way we perceive the world is influenced by our perspective as part of that living world. Imagine if we could stand back from our inborn viewpoint. Envision a visitor from another planet. Picture not the typical little green anthropoid with a third eye and a tail, but an extraterrestrial about the size of a *Salmonella* bacterium. To him, her, or it, animals, plants, and fungi, even the pinnacle of Aristotle's Great Chain of Being, the white male *Homo sapiens,* would be invisible in their enormity. But our ET would marvel at the astounding diversity of single-celled life.

Another lesson is more philosophical. The archaea remained hidden within the prokaryotes for so many years because we considered only what was obvious to our senses. We didn't have the tools and didn't know where to look or what to look for to "see" them. Perhaps the most valuable lesson that the ongoing story of Carl Woese and the archaea has taught us is that we can never know what we do not yet know. And that lesson applies to all fields of inquiry.

REFERENCES

Fox, George E., Kenneth R. Pechman, and Carl R. Woese. "Comparative cataloging of the 16S ribosomal ribonucleic acid: Molecular approach to procaryotic systematics." *International Journal of Systematic Bacteriology* 27:44–57 (1977).

Fox, George E., et al. "Classification of methanogenic bacteria by 16S ribosomal RNA characterization." *Proceedings of the National Academy of Sciences* 74:4537–4541 (1977).

Mayr, Ernst. "Two empires or three?" *Proceedings of the National Academy of Sciences* 95:9720–9723 (1998).

Morell, Virginia. "Life's last domain." *Science* 273:1043–1045 (1996).

Nisbet, Euan. "The realms of Archaean life." *Nature* 405:625–626 (2000).

Olsen, Gary J., and Carl R. Woese. "Ribosomal RNA: A key to phylogeny." *FASEB Journal* 7:113–123 (1993).

Wade, Nicholas. "Rethinking the universal ancestor: a new 'tree of life.'" *New York Times*, 13 June 2000, F1.

Wade, Nicholas. "Genetic analysis yields intimations of a primordial commune." *New York Times*, 13 June 2000, F1.

Whitman, William B., David C. Coleman, and William J. Wiebe. "Prokaryotes: The unseen majority." *Proceedings of the National Academy of Sciences* 95:6578–6583 (1998).

Woese, Carl R., and George E. Fox. "Phylogenetic structure of the prokaryotic domain: The primary kingdoms." *Proceedings of the National Academy of Sciences* 74:5088–5090 (1977).

Woese, Carl. "The universal ancestor." *Proceedings of the National Academy of Sciences* 95:6854–6859 (1998).

Woese, Carl R. "Default taxonomy: Ernst Mayr's view of the microbial world." *Proceedings of the National Academy of Sciences* 95:11043–11046 (1998).

IN PURSUIT OF PRIONS

Most stories have a clear beginning, middle, and end. In science, however, discoveries, insights, and connections sometimes go back and forth and overlap in time as pieces emerge to a puzzle that no one even knew existed. So it is for the story of prions, proteinaceous infectious particles that cause disease in a novel way.

Prions are brain proteins that can change shape and induce other proteins like them to change shape too. A prion harbors areas that can assume different forms, like a mime who can contort his or her face to take on many expressions. The shape-changing cascade seen in at least 85 species of mammals sets into motion a deadly pathology. The brain becomes riddled with holes, obliterated with masses of protein and overrun with star-shaped cells called, appropriately, astrocytes, as neurons that link the nervous and muscular systems die. At least half a dozen prion diseases are known (Table 4.1). They share certain features, such as a long incubation period, rapidly deteriorating coordination, tremors, and wasting. Yet each has a distinctive pattern of symptoms, and some have unique characteristics, such as facial numbness, itchiness, or total insomnia. All are 100% fatal.

But that's the end of the story, for now. The tale of prions began centuries ago, perhaps even farther back in time, when shepherds noticed a strange illness among their flocks.

EARLY CLUES: SCRAPIE, RIDA, AND CREUTZFELDT-JAKOB DISEASE

The disease of sheep called scrapie begins with an itching so intense that animals frantically rub against any surface, trying to scrape their skins off—hence the name. The itching progresses to tremors and then staggering. The animal wastes away, becomes blind, falls down, and dies, all within months.

TABLE 4.1

Prion Diseases in Mammals

Natural Host	Disease
Sheep	Scrapie
Goats	Caprine arthritis encephalopathy
Humans	Several variants of Creutzfeldt-Jakob disease; kuru; Gerstmann-Straussler-Scheinker disease; fatal familial and fatal sporadic insomnia
Cattle	Bovine spongiform encephalopathy
Deer and elk	Chronic wasting disease
Ranch-bred mink	Transmissible mink encephalopathy
Felines	Feline spongiform encephalopathy

Scrapie was reported in the Middle Ages, and epidemics were noted in Europe, Asia, and South Africa in the 1800s. By 1900, about 1% of the sheep in the United Kingdom had the disease. Still, even people who relished eating mutton did not become ill. In the 1930s, French researchers showed that injecting filtered brain matter from a sick sheep into a healthy sheep could pass on the illness, revealing that whatever this disease was, it was transmissible. When scrapie struck sheep on a Michigan farm in 1947 and on another midwestern farm in 1961, the U.S. Department of Agriculture sacrificed the entire flock to curtail the disease's spread.

In the spring of 1954, Bjorn Sigurdsson, a pathologist from the University of Iceland in Reykjavik, gave a series of lectures at the University of London. The lectures, entitled "Rida, a Chronic Encephalitis of Sheep," were published later that year in the *British Veterinary Journal.* Because of the obscurity of the journal and the fact that the key paper was third in a series, the world was and is largely unaware of rida, although it is an important early chapter in the ongoing story of the prion diseases.

Rida affected sheep in isolated mountain valleys in northern Iceland. Reported Sigurdsson:

> The symptoms consist of spastic movements, inco-ordinated [*sic*] gait, and tremors .... The body temperature is normal and anatomical changes outside the central nervous system have not been observed. In the advanced stages the animals do not show the normal interest in food and water and they lose condition. They move with a wobbling gait and gradually become dehydrated and emaciated. Towards the end they commonly lie on their sides with their limbs outstretched, and, although no individual muscle or group of muscles seem to be paralyzed, the animals are unable to move and usually die in a stuporous state.

The course of rida was 8 to 18 months, and it was invariably fatal. It was also contagious, which Sigurdsson observed both in the field and experimentally by injecting diseased brain extracts into healthy sheep and watching them shake and then wither. In several of these experiments, he first passed the brain material through a filter designed to keep out bacteria, concluding that "it seems therefore that the cause of the disease is a filter-passing virus."

But this was no ordinary virus. Rida—which Sigurdsson pointed out could be identical to Britain's scrapie—did not fit into the two major categories of infectious disease, acute or chronic. An acute illness, such as flu, typhus, or measles, begins quickly and follows a predictable course, and either the infectious agent or the host's immune system wins. In a chronic illness, such as syphilis or tuberculosis, Sigurdsson noted, a long battle with an irregular course ensues after a short incubation period, and the outcome of the clash with the immune system is uncertain.

Rida did not fit either description, nor did scrapie. These diseases followed a distinct and deadly course. Two other characteristics distinguished this type of disease: The incubation period was long, and the immune system uninvolved. Reported Sigurdsson, "[O]ne is even sometimes tempted to suspect that there is no effective immunity response in the slow infections .... Perhaps this infectious agent is so well adapted to its host, so well camouflaged, that it has to some extent eliminated its own species specificity in the immunological sense." The agent escapes immune surveillance, it was later learned, because it is a normal biochemical twisted into an unusual shape and because it can remain inside cells and beyond the territory of the immune system in the brain.

The term *slow virus,* which Sigurdsson originated, dominated prion research for many years. Then the evidence began to build that the infectious agent was not a virus at all, but a protein. But not many people had heard of these bizarre disorders of sheep, nor had many people heard of Creutzfeldt-Jakob disease (CJD).

The first reported case of CJD was a young maid working in a convent in Breslau, Germany, who was brought to a clinic following a rather abrupt change in habits. In June 1913, she stopped eating and cleaning, and would laugh inappropriately. She rapidly started to twitch, lost coordination, and then suffered tremors and seizures, dying three months later. A clinic doctor, Hans Gerhard Creutzfeldt, described the case in a 1920 medical journal. A physician who read a draft of the paper, Alfons Jakob, was startled to note that he had seen four similar patients, and he reported his cases the following year. On autopsy, the patients' brains showed no signs of inflammation, which would indicate the immune system's response to infection or injury, but they did show astrocytes crowding out neurons. Neither Creutzfeldt nor Jakob reported further cases, thinking that they were seeing negative evidence—areas in the brain devoid of cells, like holes in a Swiss cheese.

No one would think much about scrapie or CJD for more than three decades, when the two would become linked, however improbably, by people called the Foré (pronounced Fore-ay), who lived in the remote mountainous interior of Papau New Guinea.

OF KURU AND CANNIBALS

The second largest island in the world, New Guinea is 400 miles wide and 1500 miles long and is located in the western Pacific beneath the equator, east of Borneo and Sumatra and north of Australia (Figure 4.1). The dense interior, cut off from the rest of the world until Australian gold prospectors arrived in the 1920s, was settled by short, hefty, dark-skinned people who lived in small hamlets, several of which formed villages. About 35,000 people lived in 169 villages, which were united to form hundreds of distinct linguistic groups. Papau New Guinea was the most linguistically diverse place on earth. Like the sheep that developed rida, the Foré lived at very high elevations (3000 to 4000 meters; 9000 to 12,000 feet), a fact that may be important in the development of prion diseases.

Foré men and women lived separately. Children stayed with the women, but when boys neared puberty they were sent to a "men's house," where they would spend the rest of their lives. People married but lived apart, with spouses meeting in gardens at night to have sex. The men hunted and ate meat; the women and children ate vegetables and whatever they could scavenge, usually worms and spiders. But the Foré had another source of food intimately linked to their culture and also pragmatic—they ate their own dead. Said one oft-quoted missionary: "The group's cemeteries are their stomachs."

Anthropologists think that the Foré in the southern part of the island learned cannibalism about a century ago from visits to the northern Foré, who had practiced it for some time. The south Foré ate their dead relatives and friends to honor them, and followed a rigid set of rules for distributing the parts. Those who had been closest to the deceased would receive the best pieces. The least desirable part was the brain, often eaten undercooked; the gallbladder was the only section that was discarded, perhaps because of the bitter taste.

To prepare a body for the cannibalism rites, first the feet and hands were cut off, then flesh stripped from the limbs. The men would receive this most valued muscle tissue, which was mixed with vegetables and spices and stuffed into banana leaves, then roasted. Sometimes pig meat would be mixed into the concoction. The north Foré, in contrast,

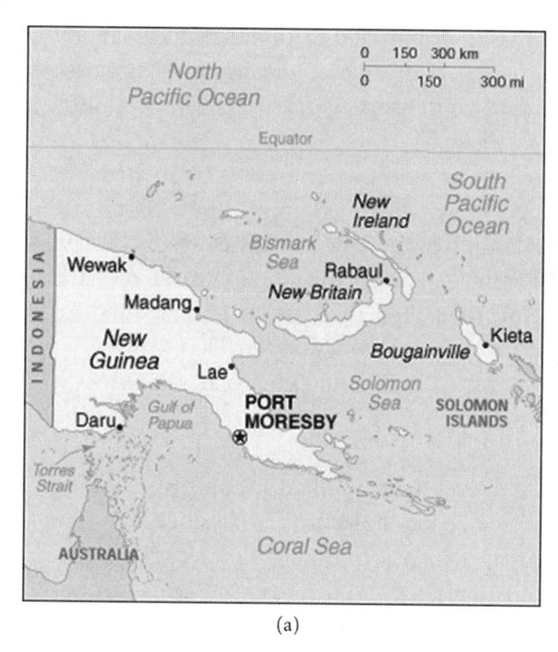

(a)

(b)

FIGURE 4.1

(a) New Guinea is an island in the western Pacific. (b) This woman with kuru is emaciated and could no longer walk or talk.
(a): Central Intelligence Agency, World Factbook, 1999. (b): Photo courtesy of Dr Robert Klitzman, Columbia University.

cooked the body whole. In the south, once the skull was opened with a swift swing of an ax, the brain would be removed and roasted in a bamboo tube. If the victim was a male, his wife would receive his sex organ. It was a great honor to be eaten. People ill with leprosy or a diarrheal disease were excluded, much to their shame, but others who knew that they were dying could indicate who they wished to receive which part.

Lean bodies were much appreciated, and by the 1950s, about 1% to 2% of the members of each village in the south died each year and provided very stripped-down corpses. These individuals were mostly women and a few children, and in some villages, the men came to outnumber the women three to one. In one village, of 125 men older than 21 years, 63 had lost wives. The Foré called the odd disease *kuru,* which means "shiver" or "tremble," and attributed it to sorcery. Richard Klitzman, a psychiatrist at Columbia University who spent time with the Foré in the early 1980s, learned to understand their pidgin English—a mixture of English, Spanish, and German—and described their own words for the illness in his book *The Trembling Mountain.*

When the initial trembling began, the person would be described as "kuru laik i-kamap nau," which means "kuru like he come up now." Within a month, coordination would begin to diminish, and the legs to wobble, a condition called "wokabaut yet," or "walk-about yet." The second month was "sindaun pinis," the ominous "sit down finish," when the trembling worsened into shakes, speech slurred, thinking slowed, and the abilities to eat and walk began to fail. Weight plummeted. The third month was "slip pinis," which means "sleep finish," as the person became mute but broke into a grimace and made sounds that resembled laughter, earning the illness the nickname "the laughing death."

Finally, the person would be deemed "klostu dai nau," which means what it sounds like. Death often came from thirst, hunger, pneumonia, or infected bedsores. Worst of all, a person suffering from kuru was often aware of his or her decline.

After death came the "time bilong kutim na kaikai," which translates as "time of cutting up and eating." D. Carleton Gajdusek, a physician who introduced the Western world to the Foré and kuru, described how women and children prepared the dead in his Nobel Prize speech in 1977:

> Children participated in both the butchery and the handling of cooked meat, rubbing their soiled hands in their armpits or hair, and elsewhere on their bodies. They rarely or never washed. Infection with the kuru virus was most probably through the cuts and abrasions of the skin or from nose picking, eye rubbing, or mucosal injury.

If whatever was causing kuru—thought to be a virus for lack of other explanations—was transmissible, it was nearly impossible for people *not* to contract the illness given the absence of sanitary conditions. Men ate cooked meat; women and children, even babies and toddlers, ate undercooked brain and smeared themselves with the raw tissue—repeatedly. Plus, cooking at the high altitude might have only partially inactivated the agent.

Gajdusek had expertise in pediatrics and virology and was just starting his career at the Walter Reed Army Institute of Research in March 1957 when he stopped on his way home from a year in Australia to visit colleagues in New Guinea. Public health officers described the strange illness that was striking the southern Foré and asked Gajdusek to help investigate its origins. The young doctor became so intrigued that he stayed, beginning what would become a lifelong commitment. After visiting several women and boys with kuru, Gajdusek began to methodically describe the illness, replacing some of the pidgin English with medicalese. The tremors became known as "athetoid movements," the end-stage grimacing "pathological laughter," a term that unfortunately led some Australian tabloids into poking fun at the disease.

The American doctor visited many hamlets, often tailed by curious children. He meticulously documented the progression of symptoms and established the epidemiology of the disease. He found that about 1% of the southern Foré were becoming ill each year, totaling some 300 or so individuals, with death tolls reaching 5% to 10% in some hamlets. About 50 of the affected people were children. The symptoms were clearly degenerative—a progressive loss of function—and the illness appeared to be infectious because several people in a village would develop it at about the same time. But, oddly, there was no fever or inflammation, the hallmarks of infectious disease.

Because members of the same families often developed kuru, it looked for a time to be inherited. However, arguing against heredity was the observation that people coming to an infected hamlet from elsewhere became ill, even though they had no blood relatives there. But some of the missionaries and public health workers who had been in the area for years, and even the 1920s gold prospectors from Australia, had talked about a link between kuru and cannibalism.

Gajdusek realized that to get to the bottom of kuru, he, or someone better qualified, would need to study affected brains, particularly the cerebellum, the region that controls coordination. Meanwhile, Gajdusek's funding had shifted to the National Institutes of Health (NIH), where a colleague, Joe Smadel, agreed that an autopsy would be crucial to publishing a description of kuru. A 50-year-old woman named Yabaiotu finally gave Gajdusek permission to conduct an autopsy—most of the Foré were afraid of it—and after her death, her brain wound up in the laboratory of neuropathologist Igor Klatzo at the NIH facility in Bethesda.

Klatzo's findings supported Gajdusek's. The brain had no sign of inflammation or infectious microorganisms, nor did it resemble the pattern of damage characteristic of such neurodegenerative disorders as Alzheimer's disease or multiple sclerosis. There were no traces of poison, and a careful analysis of everything that the Foré ate didn't reveal any natural toxins. Klatzo saw flower-shaped blackish areas in some of the children's brains that reminded him of the brains of Alzheimer's patients, but because it wasn't a consistent finding, he dismissed it. After eliminating the most obvious suspects, Klatzo noted similarity to an obscure disease described back around 1920. The brain of the woman who died of kuru had areas that looked just like the brains of a handful of people who had died of Creutzfeldt-Jakob disease. It would turn out to be an important connection.

The second connection came in 1958, when Gajdusek returned briefly to the NIH, bringing tissue samples and photos and films of people suffering from kuru. He presented seminars at universities and exhibits at museums, alerting the medical and scientific communities to what he called "a now disappearing disease in a small primitive population."

Coincidence and chance meetings sometimes propel scientific discovery, and that is what happened in the prion saga. Someone who had seen a Gajdusek kuru exhibit at a museum in London mentioned how interesting the disease was to a friend, a veterinary pathologist named William Hadlow, who had come to England from Ohio to study scrapie in sheep. Hadlow went to see the exhibit and was struck by the similarities between kuru and scrapie. The symptoms, their sequence and time course, the incidence, the ability of outsiders to develop it, the lack of fever and inflammation, and the holes in the brain were common to both disorders. Hadlow knew that scrapie could be experimentally transmitted from one susceptible animal to another. Was the same true for kuru?

Back in New Guinea, Gajdusek began assembling the pieces of evidence. Creutzfeldt-Jakob disease seemed to arise spontaneously: It was extremely rare, but present in all populations studied. Scrapie was transmissible and could be passed to minks that ate affected sheep as part of their feed (or so it was thought). Might kuru represent an initial spontaneous case of CJD that spread among humans via the unusual route of cannibalism? He too concluded that researchers needed to demonstrate the transmissibility of kuru.

INTO THE LAB TO ISOLATE THE AGENT

Gajdusek and his colleagues at NIH had been trying to pass kuru to mice, but they realized that if a "slow virus" was the culprit, the animals might not live long enough to develop symptoms. A primate model was necessary. Gajdusek began the new research project by obtaining samples of scrapie-infected brain from the British Agricultural Research Council field station at Compton in 1960, where he had gone to visit Bill Hadlow. He then set up shop at a Fish and Wildlife Service site called the Patuxent Wildlife Research Center, midway between Washington, DC, and Baltimore, where he refit several barns to house primates. Gajdusek hired virologist Joe Gibbs to try to give mice scrapie, kuru, and, for good measure, Alzheimer's disease and Parkinson's disease.

Meanwhile, Hadlow had been investigating sick ranch-raised mink. At first, mink-transmissible encephalopathy was attributed to eating scrapie-infected sheep parts in feed. But

Hadlow studied records of a 1947 outbreak that had killed 1200 mink and, lining up the dates, realized that the mink could not have gotten the disease from the sheep: Not enough time separated the outbreaks for transmission to have occurred from sheep to mink. He then linked another mink outbreak in 1963 in Blackfoot, Idaho, plus the 1947 event, to an even scarier source—"downer" cattle, cows that were obviously ill and sold to slaughterhouses cheap, destined for animal feed. The true importance of this connection would not become clear for many years.

Patuxent became the site of the first attempts to transmit kuru to primates. Chimp A-1, Daisy, was inoculated with infected brain matter on February 17, 1963, and A-4, George (later renamed Georgette when he turned out to be a she), received the same treatment. Brain matter from kuru victims was injected directly into the chimps' cerebrums. Joe Gibbs kept meticulous records on the animals. Georgette's symptoms began on June 28, 1965, with a slight shiver, an odd vacant look in her eyes, and altered behavior: She sat quietly in a corner, which was not normal for a young, active chimpanzee. Two weeks later, the tremors worsened, her jaw hung open, and she moved slowly, with poor balance. She lost her depth perception, became fatigued and skinny, and grew weaker. By October 28 she was blind. Soon afterward Georgette was euthanized, and her brain flown to Oxford University, where a scrapie brain expert named Elisabeth Beck delivered the verdict "indistinguishable from human kuru." Daisy suffered a similar fate.

From 1966 through 1968, efforts intensified on several fronts. Researchers combed missionaries' diaries to retrospectively connect specific cases of kuru to past feasts on human flesh. In 1968, a chimp injected with brain matter from a British man who had died of CJD developed CJD. Kuru, CJD, and scrapie now became firmly established as subacute transmissible spongiform encephalopathies (TSEs). (The *subacute* referred to the incubation time after infection but before symptoms appeared.) The medical community began to realize that others could be at risk, such as transplant recipients and anyone handling brain material.

The year 1968 marked a transition of sorts from clinical research to a systematic search for the TSE agent. The it-must-be-a-virus-because-it's-not-a-bacterium mindset began to change as researchers ruled out more conventional types of infectious agents. At Hammersmith Hospital in London, radiopathologist Tikvah Alper freeze-dried part of a scrapie-ridden mouse brain and bombarded it with electrons in a particle accelerator, a technique that estimates the size of the object. She found that the scrapie agent, whatever it was, was smaller than a nucleic acid, which forms the heart of a virus. Strong ultraviolet light, which would have kinked a nucleic acid strand into uselessness, had no effect on the agent. Work at the University of Edinburgh revealed that the scrapie agent seemed to come in different strains, each causing a distinctive pattern of brain holes, incubation times, and symptoms. Finally, several researchers showed that nucleases, which are enzymes that cut DNA and RNA, had no effect on the TSE agent. However, proteases, which cut protein, sometimes inactivated the infected brain matter.

It was beginning to look as though the agent that causes scrapie, CJD, and kuru was a protein. But proteins were not known to be infectious or to come in strains. J. S. Griffith, a mathematician at Bedford College in London, published a paper in the scientific journal *Nature* in 1967 suggesting several ways that a protein could assume these novel roles. One of his ideas would ultimately gain credence—namely, that the TSE agent was an abnormal conformation of a normal protein that forms spontaneously and then bends other normal copies out of shape. Such a phenomenon would explain the infectivity and the mysteriously absent immune response.

And so from 1957 until about 1970, research into the TSEs gradually focused in on the cause. The research began with epidemiology in New Guinea and became clinical as Gajdusek and his coworkers described the disease. The effort then became microbiological, with attempts to inoculate other species to demonstrate transmissibility. Physical and biochemical testing came next, which ruled out a nucleic acid and pointed to a protein. And at the close of the decade, the inquiry was turning theoretical, as investigators finally distanced themselves from the idea that only a microorganism or virus could cause infectious illness.

Ironically, as the biomedical community began to close in on the TSE agent, kuru was disappearing among the south Foré. It vanished first among the very young, in the wake of cessation of cannibalism. Kuru was now affecting only the elders. Meanwhile, the Foré began to cultivate coffee and communicate with the outside world. But the world certainly hadn't heard the last of the TSEs.

IATROGENIC CREUTZFELDT-JAKOB DISEASE ARISES

The cases of CJD reported in the 1920s occurred spontaneously—that is, out of nowhere, without apparent cause. Rarely, CJD affected families, where it was caused by a gene mutation. In the 1970s, however, scattered reports of the disease linked to tissue and organ transplants began to appear in the medical literature. These cases were *iatrogenic,* meaning "physician caused."

The first case was a middle-aged woman who, in 1971, went to an ophthalmologist at Columbia University complaining of cloudy vision. The physician properly diagnosed Fuchs dystrophy and treated it with a corneal transplant, donated from a man who had died of pneumonia following two months of undiagnosed tremors and memory loss. Although the transplant helped the woman's vision, a year and a half later she developed very disturbing symptoms: She drooled, couldn't keep her balance, and lost a great deal of weight. She entered a vegetative state and died. Upon autopsy, her brain had the characteristic CJD holes. The infectious agent had traveled from the grafted cornea to the woman's brain, probably via the optic nerve.

Years later, the donor and recipient's tissues were reevaluated and passaged through chimps, who became ill with CJD. But not everyone accepted the diagnosis. "The very thought of human-to-human transmission of a degenerative disease by physicians was so unspeakable that some detractors suggested that the presence of disease in the donor and the recipient might have been coincidental," wrote Richard Johnson of Johns Hopkins Hospital and Clarence Gibbs Jr. of the NIH in the *New England Journal of Medicine* in 1999. Both were involved in the initial corneal transplant case.

Then in 1973, Gajdusek and Joe Gibbs reported on a middle-aged male neurosurgeon who had died of melanoma but also had symptoms that had been diagnosed as Alzheimer's disease. Again, samples of the neurosurgeon's brain implanted into chimps' brains caused CJD. This was case number 2, but the medical community was still not convinced that CJD could be transmitted by transplant or occupational exposure to infected tissue.

The next two cases nailed the argument and strengthened warnings that those who handle brain tissue should be careful. A Swiss physician reported a 23-year-old woman and a 17-year-old boy who developed CJD from contaminated silver electrodes placed in their brains to identify locations for surgery to relieve severe seizures. Eventually the electrodes were traced back to their use in a 69-year-old woman who had died of CJD in 1974.

Cleaning the electrodes in formaldehyde and alcohol had not destroyed the agent. Both brains were full of holes. And material from both brains transmitted the condition to chimps.

From 1977 until 1982, a mini-panic set in. The *New England Journal of Medicine* published warnings and guidelines for working with brain and for identifying patients who might have CJD and should not donate tissue. Experiments with guinea pigs showed that the agent could be passed in a blood transfusion and that various tissues, including spleen and kidney, could transmit kuru, CJD, and scrapie to monkeys. Scrapie brain rubbed into mouse gums spread the illness, revealing a dental route. In England a dentist and two patients developed CJD, possibly through gum exposure.

Despite the precautions that seemed to work well through the 1980s, cases of iatrogenic CJD still appear, many likely misdiagnosed as Alzheimer's disease or other more common neurodegenerative conditions. For example, a 53-year-old woman who died of lung cancer in early 1997 donated her corneas to two people. When the woman's daughter mentioned that before death her mother was "acting like a senile old lady" and had lost coordination, at first doctors attributed this to spread of the cancer to her brain. But an autopsy months later revealed the hole pattern of CJD. The recipients had to give their corneas back. Considering the tens of thousands of corneal transplants performed each year and the fact that CJD often masquerades as other disorders, the likelihood of a few iatrogenic cases per year remains with us. Gajdusek calls this route to a TSE "high-tech neocannibalism."

HUMAN GROWTH HORMONE FROM CADAVERS CAUSES A PRION DISEASE

Yet another unusual group of individuals would soon prove vulnerable to the still-unknown TSE agent. Lack of growth hormone (GH), which is produced by the pituitary gland in the brain, causes severe dwarfism. In 1958, researchers found that replacing GH with the hormone pooled from cadavers enabled children with this condition to grow. In 1963, the NIH began the National Hormone and Pituitary Program to collect the material and test its efficacy in affected children. Over the years, GH treatment enabled many children to grow taller than their mutant genes dictated. But a problem emerged. In late 1976, aware of the iatrogenic cases of CJD, geneticist Alan Dickinson at the Institute for Animal Health in Edinburgh alerted the British government and his colleagues to the possibility of spreading CJD through cadaver GH. It was a terrifying scenario, because thousands of glands were used to produce batches of the hormone.

For the most part, Dickinson's warning was ignored. Sadly, he was right. In 1984, a 21-year-old man who suffered from several hormone deficiencies and had received cadaver GH since early childhood suddenly became dizzy on a trip to visit relatives. He rapidly lost his coordination, remained dizzy, and began twitching. Six months later he died of CJD. His doctor, Raymond Hintz at Stanford University, reported the case to the Food and Drug Administration and NIH, and the agencies responded by restricting use of the cadaver hormone to only the most severely affected patients. While study of the situation continued, some physicians objected to such a wide ruling based on only one reported case. But then several more cases were noted, originally misdiagnosed because the victims had been too young to fit the typical CJD profile, and kuru had seemed out of the question. CJD in these young people fit the symptom pattern of kuru more than it did that of CJD, with loss of balance predominant rather than the dementia associated with CJD. As the number of cases increased, the NIH was forced to stop the cadaver GH program.

Fortunately, a plentiful and pure source of the hormone soon became available thanks to recombinant DNA technology, and a potential GH–prion disease disaster was averted.

Scattered cases of the cadaver-derived TSE appeared in the 1980s. But in retrospect, a few of the cases from Britain near the end of the decade might have had another cause—contaminated beef.

PRUSINER AND PRIONS

J. Carleton Gajdusek received an early-morning phone call on October 14, 1976, informing him that he would share in the Nobel Prize for physiology or medicine for his work on kuru. Two decades later, another researcher would receive the prize for work on the same problem. He is Stanley Prusiner, a professor of neurology, virology, and biochemistry at the University of California, San Francisco (UCSF).

Prusiner became interested in the TSEs in 1972, when, as a neurology resident at UCSF, he lost a patient to CJD. In 1974 he obtained funds from the university to pursue identification of the disease-causing agent, but applied to NIH for additional support because investigating a long-term disease in mammals is costly. In the classic catch-22 scenario of the young scientist, NIH informed Prusiner that he couldn't receive a grant to study something unless he could prove that he already had experience studying that something—which, of course, he needed funds to do. The committee that reviewed his grant proposal suggested that he learn virology, so he went to Sweden for several years to do that, and then spent time at the Rocky Mountain Lab with scrapie expert Bill Hadlow. By 1978 Prusiner was back at UCSF, traveling that year and again in 1980 to visit New Guinea and work with Gajdusek on the clinical aspects of kuru.

Back in the UCSF lab, Prusiner found unbearable the months he had to wait for mice to die of scrapie so that he could study them. He devised a more efficient system. He used hamsters, which sicken much faster, and examined them at the first sign of illness, not after death. Prusiner and his colleagues used chemical methods to enrich hamster brain extracts for the mysterious infectious agent, a little like boiling and straining tomatoes to make a concentrated sauce, searching for a chemical present in the brains of infected animals but not in healthy ones. They repeated Tikvah Alper's finding that the culprit was not a nucleic acid, as well as performed complementary experiments showing first that it was protein, and eventually that it was a single type of protein.

The causative agent of the TSEs remained nameless, still known for what it wasn't—a nucleic acid—than for what it was. Gajdusek wanted to wait until they knew its chemical identity to name it, but Prusiner, in a 1982 paper in *Science,* dubbed it a "prion" (pronounced pree-on), a short version of "proteinaceous infectious particle." Gajdusek was not pleased.

Although Prusiner hedged his bets by writing that it was possible that a prion was sequestering a small nucleic acid that could not be detected, he soon became strongly linked with the "protein-only" view of disease transmission. It wasn't a very popular idea. "When Stan proposed the protein-only prion hypothesis and stuck with that, it was at a time when it was known that infectious diseases result from particles with nucleic acids, and it was dogma that a protein sequence folded into a single structure. In recent years, it has become clear that proteins are not necessarily such static entities with well-defined structures," says Thomas James, chair of the department of pharmacological chemistry at UCSF, whose recent work on imaging prions has clarified much of their function. Biochemists have since learned that some proteins can assume different forms transiently. Prions stand out in that they represent a protein that comes in more than one *stable* form.

Prusiner's research group grew, and experimental results poured out. In 1982 they isolated a protein of a certain size that was found only in brains of scrapie-ridden animals and could not be digested by proteases as most proteins are. In 1983, they named the protein PrP, for "prion protein," and described it as aggregating rods consisting of up to 1000 PrP particles. But someone else had already visualized PrP. In the late 1970s, a predoctoral researcher at the New York State Institute for Basic Research in Developmental Disabilities on Staten Island, New York, perfected a method of imaging scrapie-infected hamster and mouse brain cells in the electron microscope. Patricia Merz saw, over and over again, piles of matchsticklike structures in the scrapie preparations but not in the normal controls. What's more, the number of matchsticks increased with the severity of illness, and the different strains of scrapie were associated with slightly different-appearing matchsticks. Back then Merz had named the mysterious matchsticks "scrapie-associated fibrils" (SAFs) and went on to find them not only in the brains of mice and hamsters with CJD but also in their spleens. Unsure of herself, Merz for years thought that what she was seeing was just another manifestation of the disease, not the causative agent. Some researchers insist that she, and not Prusiner, initially identified prions.

By 1984, several research groups had seen SAF/PrP in brains of animals who had died of scrapie, kuru, or CJD, but not in brains of people who had died of Alzheimer's disease, Parkinson's disease, or amyotrophic lateral sclerosis (ALS). It was beginning to appear that the so-called slow virus diseases—scrapie, kuru, and CJD—were actually prion diseases.

Also in 1984 came a torrent of biochemical information. Prusiner and Leroy Hood's group at the California Institute of Technology discovered a distinctive stretch of 15 amino acids at one end of PrP. They used the information in its sequence to synthesize a DNA probe, which, when tagged with a fluorescent dye and placed on a chromosome preparation, homed in on the corresponding gene. Several research groups used the approach to quickly identify genes encoding PrP in hamsters, mice, humans, and eventually many other mammalian species. Most astonishing was that researchers consistently detected this gene, and its protein product, in normal cells! As unexpected as this result was, it did clear up one part of the mystery—it explained why there was no immune response to kuru, CJD, or scrapie. The body does not attack what it regards as part of itself.

If a gene causes symptoms in one guise but not another, it is easy to explain: The gene mutates into another form, encoding a slightly different amino acid sequence for the corresponding protein. But this simple explanation did not turn out to apply to the TSEs. The PrP in normal and scrapie brain cells had the same amino acid sequence. However, the PrP particles from the two brain types did differ. Normal PrP, like most proteins, falls apart in the presence of a protease; abnormal PrP proved stubbornly resistant to protease. Prusiner now distinguished two types of PrP: "cellular PrP," which did not cause disease, and "scrapie PrP," which came to represent any TSE-causing variant of the protein. At this point, the research veered from the biochemical to the theoretical.

A gene's nucleotide base sequence dictates a corresponding protein's amino acid sequence. As a protein is made, it folds into a characteristic three-dimensional shape, or conformation, that is essential for its normal function. Could PrP sometimes fold improperly, generating an abnormal form that somehow sets off the chain reaction that is a TSE? If this was the case, then the genes for both forms would be the same. Plus, there could indeed be different "strains"—they would simply be different conformations of the same amino acid sequence.

But how does the abnormal form of PrP arise? Prusiner hypothesized that somehow the "evil twin" prion converts, or subverts, the "good twin" into the pathogenic form. They already had abundant evidence for this: Injecting infected brain matter into healthy animals spreads the illness, hence the word *transmissible* in "transmissible spongiform encephalopathy." The question of how the first abnormal PrP arises, however, remains unanswered. It may simply happen by chance—a rare, although natural, event. A protein may fold one way 99.99% of the time, and the fleeting alternate way yields the prion form that can then go on to cause widespread destruction.

A changeable form of a protein could explain how TSEs could be both infectious and inherited. A small percentage of cases of CJD and a few other rare neurodegenerative diseases are caused by mutations in the PrP gene that alter the protein in a way that produces a shape like that of the scrapie form. Prusiner's group studied PrP mutations, starting with a man who had inherited the rare Gerstmann-Straussler-Scheinker disease. He had a very distinctive mutation in the PrP gene. The researchers used this information to genetically engineer a type of transgenic mouse that carried the mutation in all its cells. As expected, these mice developed spongiform brains; most importantly, extracts of their brains could transmit the illness to healthy, nontransgenic mice. This showed that once a prion becomes abnormal, its abnormality is transmissible. (Gajdusek and others had shown in the early 1980s that brain material from people with the rare inherited prion diseases could transmit the illness to monkeys and mice.)

Prusiner and his coworkers went on to create various other types of transgenic mice, which provided compelling new clues to the nature of the TSEs, as well as tools to study them. Mice altered to produce hamster cellular PrP quickly became ill after being inoculated with brain matter from hamsters with scrapie. Because normal mice do not contract scrapie from hamsters, these results indicated that a scrapie PrP can trigger conversion of cellular PrP if the host's PrP is similar to the infecting PrP. Mouse and hamster PrP differ in 16 of their 254 amino acids, but transgenic mice, because they have hamster PrP genes, were a perfect match. Most interesting were mice that made both mouse cellular PrP and hamster cellular PrP. When given infected mouse brain, these double-duty mice made mouse prions, but when given infected hamster brain, they made hamster prions. Somehow PrPs from the same species can recognize one another.

The transgenic mouse studies revealed that TSEs do not honor the sharp species barriers that many infectious diseases do. That is, a prion disease can jump from one species to another if their PrP genes are similar in sequence. Sheep and cattle PrP, for example, differ in only seven amino acids, a fact that would soon hit home in England.

THE BRITISH BEEF SCARE

Shepherds had seen sheep afflicted with scrapie for centuries. Missionaries in New Guinea had known about kuru for decades, and CJD had first been described in the 1920s. The emergence of the next prion disease would be much faster, developing on a weekly basis and sending meat-loving Britons into a state of panic. With the epidemic of bovine spongiform encephalopathy (BSE, or "mad cow disease") and the outbreak of "new variant CJD" (nvCJD) among humans that followed, investigation of the prion diseases transcended biomedical science and public health to embrace politics and the press. It was a scary time.

1985 TO 1987: 420 NEW BSE CASES

The first cow to display the odd symptoms of BSE so alarmed the veterinarian who examined her that he remembers the date—April 25, 1985. The black-and-white Holstein, from a farm in southwest England, had become nervous and skittish, and then began falling down for no apparent reason. Probably about 100 cows became ill that year, but were not reported to the Ministry of Agriculture, Fisheries and Food (MAFF).

In November 1986, a farmer who had a few cows ill with what was obviously a neurologic disorder killed the animals and insisted that the brains be studied. When veterinarians at the Central Veterinary Laboratory in Weybridge examined the brains, they noted a spongy appearance and sent material to Patricia Merz. By early 1987, the verdict was in: The cow brain had the telltale SAFs that indicate prions. The October 1987 report in the *Veterinary Record* described this first case and coined the term *bovine spongiform encephalopathy.* The report noted that the brain did not resemble a scrapie or kuru brain in that it had more plaques than spongy areas; therefore, it looked like it was a new disease. This extremely important finding would come back to haunt the British government a decade later.

Meanwhile, MAFF had begun studying the epidemiology of the new disease, scrutinizing 200 herds and setting up a committee to find ways to stop the spread. The most obvious explanation was scrapie-infected sheep meat in the cattle feed, assuming that the cow disease was actually scrapie. By the end of 1987, although 420 cases of BSE were reported, no one was particularly alarmed. A few political cartoons even poked fun at the tottering bovines.

But epidemiologists were busy doing what epidemiologists do—trying to determine just what the infected herds did and did not have in common. All breeds were affected, so the new disease didn't seem to be genetic. And cows were becoming ill all over England, so it wasn't a toxin confined to one area. The cases were appearing all around the same time, suggesting some shared experience. When the researchers realized that dairy cows were affected much more than cattle used for beef, they had their first clue—feed.

Dairy cattle need extra protein to enable them to produce abundant, high-quality milk. Because Britain doesn't grow much soybean, a typical source of supplemental protein, farmers used meat-and-bone meal in feed. Sometimes calves ate it, and sometimes beef cattle ate it when slaughter time neared, but it was predominantly a staple of the dairy cow diet. Dairy cows were usually slaughtered for their meat when their milking days were over, typically ending up in meat pies and hamburger, rather than as steaks, because their older meat was tougher.

Meat-and-bone meal originates in a hideous place called a knacker's yard, which is where slaughterhouse leftovers, as well as cattle and sheep too ill to have made it to a slaughterhouse at all, end up. The animal hunks—mostly intestines, heads, blood, and tails, all together called *offal*—are rendered and the fat removed, and the remaining grainy material is processed into cakes called greaves, which are then ground into meal, powder, or pellets. Whatever the final form, meat-and-bone meal stinks of blood and excrement, which could be somewhat masked by mixing it with vegetable matter to produce feed for dairy cows.

The pieces of the puzzle were there, but it was mostly scientists who were aware of them. BSE was a prion disease, and, as Gajdusek and Prusiner had shown, these diseases can be transmitted through food.

1988: 3072 NEW BSE CASES

1988 was the year of the feed ban, and of what some would call the British government's downplaying and outright deception concerning the danger BSE potentially presented to human health. Epidemiologists continued to study the diets of cattle, and feed suppliers confirmed the grisly contents of the feed. In April, the government appointed Sir Richard Southwood, a retired zoologist from Oxford University, to chair a four-person committee to provide advice. None of the members was an expert on the spongiform encephalopathies, and by this time, the government knew that this was what it was dealing with. The committee soon released the Southwood Working Party Statement, which presented the following logic: The sick cows had scrapie; people do not get scrapie; therefore, people could not get sick from cows. The fallacy was that no one had shown that what the cows had was in fact scrapie—this was assumed.

In July, in response to the Southwood report, the government banned sheep and cow parts from cow feed. This was supposedly done to protect the cows, not humans. MAFF required that farmers turn in all affected cows; they would be paid half the value of a healthy cow. MAFF would slaughter the sick cows, and the Southwood committee reassured the public that this would stem the spread of disease. But no one regulated the farmers.

A report in the *Veterinary Record* in October showed that BSE could be transmitted to mice, and in December, the disease was officially declared a zoonose, that is, a disease that passes from one species to another. A group of veterinarians familiar with the other TSEs used the mouse incubation period of 10 months for BSE to extrapolate that the cows must have been infected in the winter of 1981 and 1982. That figure ushered in a new investigation of the meat-and-bone meal manufacturing process. Had anything changed at around that time?

Government veterinarians found two changes. First, rendering plants had stopped using certain solvents to extract fat because the solvents had exploded in one plant, causing a devastating fire. Fat had been extracted to extend the shelf life of the greaves. When solvent use fell, the fat content of greaves, and therefore of the meat-and-bone meal made from it, more than doubled, to 12%. The grave significance of this fact is that it was known from kuru that prions concentrate in brain tissue, where neurons are wrapped in fatty sheaths. The second change was institution of an American method of continuously rendering fat at a lower temperature than the standard British high-temperature batch method. The temperature shift was an economic measure, instituted at a time of cost consciousness and containment. More fat and lower temperature were apparently good news for the prions, just as the high altitude may have been among Icelandic sheep with rida and among the Foré who developed kuru. Epidemiologic evidence soon emerged of a link. Two plants in Scotland were the last to switch to the new techniques, and they were the last two places to report BSE.

By the end of 1988, the recipe for disaster was already written. A series of errors and oversights laid the groundwork for the scare to come:

- The government did not reimburse farmers for their potentially tainted meat-and-bone meal. Stuck with the stuff, many farmers fed it to their cattle.
- MAFF required destruction of obviously sick cattle. But a prion disease incubates for years before symptoms appear. If the disease was spread horizontally, from cow to cow, or vertically, from cow to calf, then there were plenty of beasts

quietly incubating the illness on the fields of England. And people were surely eating prion-infected beef. How many were in the subclinical stage? How many still are?

- The government still assumed that the cows had scrapie, despite the fact that cows had lived with scrapie-ridden sheep herds for centuries without getting sick.
- Paying farmers half-price for sick cows was hardly an enticement to cooperate. Many farmers, it was learned later, sent to slaughter younger healthy cows, or those showing the very first signs of illness. This way they would receive the full market value for their animals.

Amidst the government's denials, 1988 did offer an important warning—an article in June's *British Medical Journal* stated that humans could contract a prion disease from another animal species. And so battle lines were drawn. Prion researchers, familiar with the literature on CJD, scrapie, and kuru, insisted that prion diseases can jump species boundaries. But MAFF, concerned with preserving public calm, not to mention the beef industry, was quick to announce that this was impossible.

1989: 7627 NEW BSE CASES

The year 1989 brought conflicting messages from the government. At the same time that the MAFF Chief Veterinary Officer went on the BBC to repeat the mantra that BSE was scrapie and people don't get scrapie, the agency nonetheless took measures to stop the spread. In July they banned export of cattle born before July 18, 1988, and of the offspring of sick animals. In November, they instituted a "specified offals ban," an action that would prove to be a turning point in the BSE story. Specified offals were a distinct list of parts that ended up in the knacker's yard: brain, spleen, thymus, spinal cord, intestines, and tonsils.

The numbers of sick cows continued to climb, and a new type of victim appeared—a few zoo animals became ill from feed. Still, the Southwood Committee sent reassuring messages. Stated a news release, "Scrapie has been endemic in Great Britain for centuries without there being any evidence to show an incidence of CJD higher than the international average." Continuing their not-to-worry stance, the committee predicted that cows would continue to become ill until the epidemic peaked in 1992, that cases would plummet by 1996, and that altogether, 17,000 to 20,000 animals would be lost. They still thought BSE was scrapie, and they announced that cattle would turn out to be a "dead-end host." The risk to humans, they concluded, was remote.

The Southwood report had scientific flaws. Their assessment began with faulty assumptions and used them to reach predetermined conclusions even in the face of conflicting data mentioned in the very same report, such as the passage of BSE to mice. The government was also guilty of not replicating results, a key part of scientific investigation. But the public rarely pays much attention to scientific or government reports. So when a prominent neurologist announced on the BBC that his family had stopped eating beef, the press belatedly woke up, as the government quietly advised researchers not to speak publicly about the risk.

1990: 14,371 NEW BSE CASES

The new decade began with more government denial. A document called the Tyrrell report stated "Many extensive epidemiological studies around the world have contributed to the current consensus view that scrapie is not causally linked with CJD. It is urgent that the same reassurance can be given about the lack of effect of BSE on human health." It did not state that there was such a lack of effect. Plus, evidence was still nonexistent that BSE and scrapie were the same disease. The Tyrrell recommendation took a shockingly passive stance: Monitor new cases of CJD to detect any increase in incidence. The National CJD Surveillance Unit in Edinburgh reopened tracking, which had been discontinued because of the extreme rarity of CJD in any guise.

Meanwhile, BSE was showing a predilection for other species. In zoos, five types of antelopes contracted the illness from feed: a nyala, a gemsbok, a greater kudu, an eland, and an Arabian oryx. When experiments demonstrated transmission from cow to cow via intravenous delivery or brain implants, and orally to mice, the government still downplayed any danger, stating, "These results demonstrate that the disease can be transmitted using unnatural methods of infection, which can only be done experimentally in laboratory conditions and which would never happen in the field." In science, one never says never. The European Union wasn't buying the snow job. They further restricted exports, limiting calves to only those under 6 months of age and banning export of certain parts of the animals not used for food.

In May, the nation panicked when a housecat named Max died of what looked like BSE. The symptoms were classic—he began to fall asleep while standing, and staggered, twitched, and itched. His death ushered in a new slew of agricultural regulations; some schools temporarily banished beef from the cafeteria.

In June, a physician and microbiologist from the University of Leeds, Richard Lacey, blasted the Southwood conclusions and the government's actions and inactions before Parliament's Agriculture Committee. He pointed out that banning offals wasn't enough to halt the spread of infection, because other organs could carry prions in lymphatic vessels and nerves. He emphasized the lack of evidence that BSE is scrapie. To counter Lacey's accusations, Minister for Agriculture John Gummer was prominently photographed feeding his young daughter a hamburger outside Parliament, winding up, of course, on page one of the tabloids. Then he told the National Consumer Council that there was no risk of contracting BSE by eating beef. In science and medicine, of course, there is never "no risk." The evening television news continued to show staggering cattle, however.

When a laboratory pig and a marmoset died of BSE in September, the government banned use of offals in food for pets and laboratory animals. They upped the price paid to farmers to turn in sick animals. Public reaction intensified. The Germans banned British beef, Lacey advised slaughtering infected herds, more schools served beefless meals, and people began to eat more poultry and vegetables.

1991 TO 1992: THE NUMBERS OF NEW BSE CASES PEAK AT 25,644 AND 36,924

The years 1991 and 1992, when the BSE epidemic peaked, were the calm before the storm, because humans had not yet become ill (Figure 4.2). Early in 1991, BSE began to appear in calves born after the feed ban, indicating passage from the mother or some other way. Fearing alternate routes of transmission, the government banned use of meat-and-bone meal in fertilizer. In 1992, more regulations prohibited transport of cow embryos and outlawed removal of cow brains in an area where food is present. Then three cheetahs and a puma died in zoos. The *British Medical Journal,* which would chronicle the rise of the new prion illness in humans, editorialized that there was insufficient evidence for the government to assure public safety.

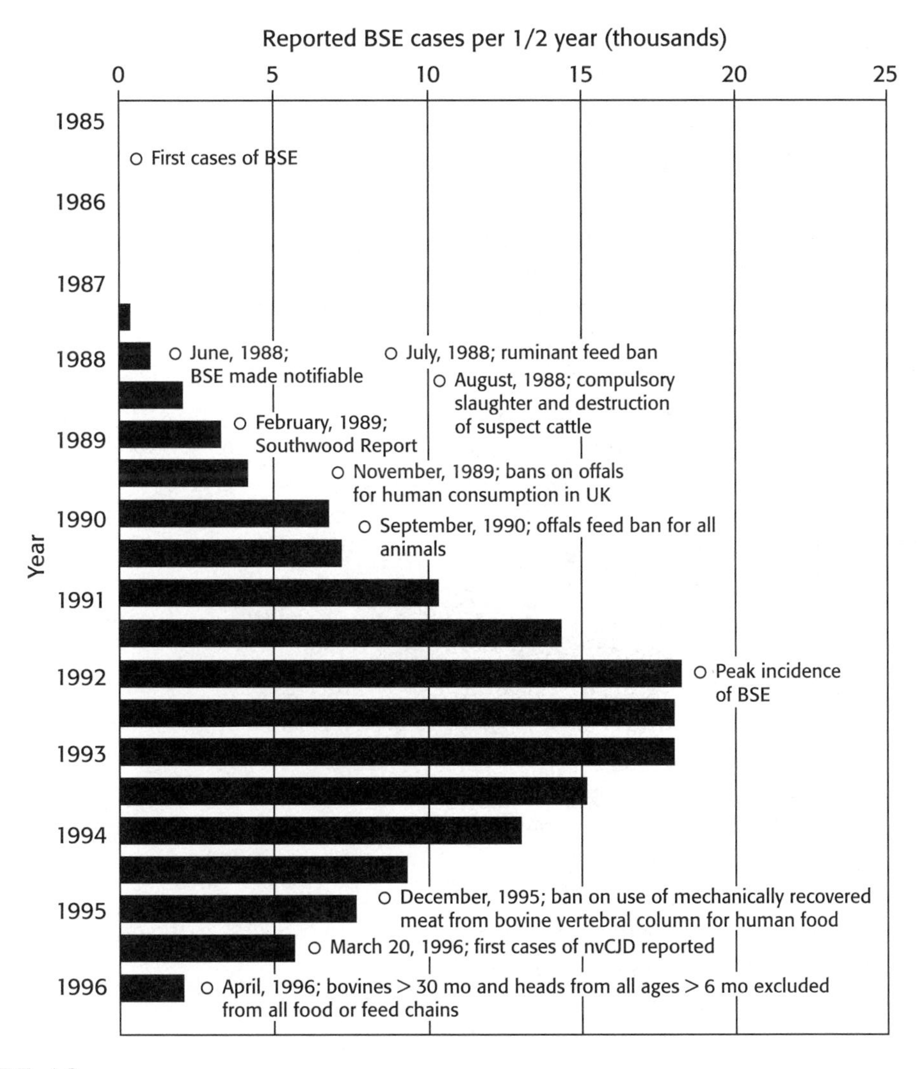

FIGURE 4.2

Evolution of an epidemic. Cases of BSE peaked in 1993.
Adapted by permission from Nature *(Anderson RM, et al. Transmission dynamics and epidemiology of BSE in British cattle.* Nature *1996;382:979), copyright 1996 Macmillan Magazines, Ltd.*

Also in 1991, an important paper by John Collinge, Mark Palmer, and Aidan Dryden of the Prion Disease Group at St. Mary's Hospital in London appeared in *The Lancet,* although mostly only others in the field took note, perhaps because it did not deal with BSE. They examined the prion proteins of the 6 people of the 1908 individuals in the United Kingdom exposed to pooled cadaver human growth hormone who had developed CJD. Four of the six had a peculiarity at the 129th amino acid position in the protein: All their proteins had the amino acid valine at this site. It would be learned later that all valines or all methionines at this site in PrP raise the risk of destablizing the protein. In contrast, individuals with methionine at this position in some of the proteins and valine in others have a much decreased risk of contracting a prion disease. Having two different amino acids at the same site indicates that the person is a heterozygote, or carrier, with slightly different sequences for the gene on the two chromosomes that carry it.

1993: THE NUMBER OF NEW BSE CASES BEGINS TO FALL (33,574)

By early 1992, 1000 cows per week were being stricken, and the 100,000th case was reported. The Southwood Committee's prediction of a peak in 1992–93 would prove accurate, but their assessment of 20,000 total cases was considerably off the mark. Two dairy farmers who had BSE-afflicted cows, Mark Duncan and Peter Warhurst, died of CJD. But because farmers and ranchers were known to have a higher incidence of CJD than the general public, their cases were attributed to sporadic CJD, not a disease arising from BSE.

On the basic research front, Charles Weissmann, a molecular biologist at the University of Zurich, developed a "knockout" mouse that was genetically engineered to lack the PrP gene. The knockout mice did not get scrapie, indicating that PrP in some form must be present to contract a TSE. They also demonstrated that whatever normal PrP does, it apparently isn't vital for survival, at least during the 70-week life span of lab mice (compared with 100 weeks in the wild, assuming escape from predation). The PrP-less mice did show a few irregularities—low levels of an inhibitory neurotransmitter, impaired long-term memory and circadian rhythms, poor coordination as they aged, and shortened life spans—but they appeared otherwise healthy, happy, and fertile.

The most striking event of 1993, and perhaps of the entire BSE saga, was the sad case of Victoria Rimmer, the first official victim of new variant CJD. In May, the spunky 15-year-old freckled blonde, who loved dogs and horses and lived in the countryside with her grandparents, began coming home from school and collapsing into a deep sleep. She became clumsy, staggered about, and couldn't remember things. Then her eyesight failed and she lost weight. Her grandmother Beryl at first thought the girl had anorexia nervosa, but by August Vicky was so ill that she had to be hospitalized. The teen couldn't talk, walk, see, move, or swallow.

After eight doctors couldn't diagnose the condition, one finally suggested CJD, which a brain biopsy confirmed. When Beryl received the grim diagnosis of "spongiform encephalopathy," she didn't know what that meant. Another doctor blurted out that it was mad cow disease, which Beryl, like most Britons, knew about from the tottering bovines on the nightly news. Convinced that eating hamburgers had sickened Vicky, Beryl wanted to speak publicly about the experience, but the researchers in Edinburgh urged her to remain silent, fearing collapse of the beef market and public panic. Unlike most human victims of prion diseases, Vicky outlived other young people who would develop the disease after she did, dying in 1996. Some researchers have suggested that she lived longer because, as perhaps the earliest known case, she received a lower dose of the agent. Laboratory experiments show that animals given greater exposure to scrapie PrP develop symptoms sooner. However, the effects of dose and duration are still not understood.

1994

Regulations were tightened: Exported cattle had to be from herds that had not had BSE for the past six years. Spongiform encephalopathies in any animal were now reportable events, and mammalian protein was banned from all animal feed. The number of cases of BSE still rose, however, to a total of 120,000 cases.

Then two more teens developed CJD. In March, a 16-year-old reported pain when his face and fingers were touched, backaches, and frequent falling. By August, his speech had slurred and his balance and coordination worsened. A brain biopsy revealed plaques, and an antibody test showed abnormal PrP. The third case, an 18-year-old man, experienced a different course of mostly psychiatric symptoms, such as sudden depression and loss of memory. As he put it, he had "just gone nutty." He soon developed delusions, hallucinations, and irrational fears. Pain or numbness to the touch and drastic personality changes would turn out to be hallmarks of the new disease, both signs pointing to a damaged nervous system.

1995

While Vicky Rimmer lay in a coma and the other two teens died, seven other young people became ill. It was undeniable: The 1988 feed ban and 1989 offals ban had been too-little-too-late actions.

The editors of the *British Medical Journal* rounded up experts for the November issue whose opinions, run under the banner headline "Creutzfeldt-Jakob Disease and Bovine Spongiform Encephalopathy: Any Connection?" revealed disagreement on the severity of the problem. The experts seemed to pick and choose the experimental and epidemiologic facts to consider. Most dismissed the cases of the farmers because a few farmers developed CJD each year in other European nations as well. Most were alarmed by the affected teens because the average age of a person with sporadic CJD is 65. Some sided with the mounting scientific evidence, others with the government's effort to dampen panic.

Jeffrey Almond, a professor of animal and microbial sciences at the University of Reading, pointed out that the PrP gene in cows is closer in nucleotide sequence to that of humans than it is to that of sheep. This meant that if the disease was in fact scrapie, it could jump from cow to person. Almond suggested two experiments: infecting mice that lack the PrP gene with BSE brain matter, and comparing the pattern of brain lesions from the teens to that of cows. If BSE and nvCJD turned out to be unlike other prion diseases and like each other, similar brain lesions in cows and the teens would be powerful, albeit indirect, evidence that the human disease arose from the bovine one. The experiments would eventually be done.

The opinion of Paul Brown, the medical director of the U.S. Public Health Service, contrasted with Almond's suggestions of specific experiments. Brown wrote that it is "most unlikely that infection with bovine spongiform encephalopathy will ever be proved to have caused any individual case of CJD." He also dismissed the need for further meetings, "because the precautions taken some years ago to eliminate potentially infectious products from commercial distribution were both logical and thorough."

But Sheila Gore of the Institute of Public Health in Cambridge countered Brown, claiming that the new cases of CJD were "more than happenstance," deeming the cases an epidemiologic alert. Flip-flopping again, R. M. Ridley and H. F. Baker, of the School of Clinical Veterinary Medicine in Cambridge, offered a litany of rationalizations and alternate explanations, such as multiple coincidental cases of inherited CJD, or a variant of sporadic CJD that had never been seen before or had been misdiagnosed.

Kenneth Tyler of the VA Medical Center in Denver provided a microbiologist's perspective. He identified the amount of material required for infection, the route, and host susceptibility. He pointed out that the tissues that are most heavily infected are not the ones that are typically eaten, and that oral delivery is possible but not likely. The prions would have to weather the acidic vat of digestive enzymes in the stomach, then enter the bloodstream and traverse the blood-brain barrier. Tyler brought up the 1991 finding that key differences in PrP can affect susceptibility to illness.

1996

Despite the *British Medical Journal*'s panel of experts, much of the medical profession remained unaware of the new disease. When Peter Hall, a freshman at Sunderland University, developed symptoms identical to those of Vicky Rimmer, his parents specifically asked if he might have CJD, but their neurologist said it was impossible because he was too young. Fifty other physicians agreed! Peter died in February 1996, after a devastating illness that lasted 13 months.

On April 6, *The Lancet* published a paper from R. G. Will and his coworkers at the National CJD Surveillance Unit in Edinburgh acknowledging that a new disease was striking young people in England and that it might be the consequence of the BSE epidemic. The paper, "A New Variant of Creutzfeldt-Jakob Disease in the United Kingdom," appeared nearly 11 years to the day that a veterinarian reported the first sick cow. Anticipating a press panic, British health secretary Stephen Dorrell announced the results on March 20 of the examination and comparison of 10 cases of CJD in Britons younger than 42 years. Dorrell had earlier insisted that there was no possibility of people becoming ill from eating British beef.

The 10 cases of new variant CJD stood out among the 207 CJD cases that the team had tracked since reconvening in 1990. The researchers evoked the term "new variant" for three compelling reasons:

1. The victims were young.
2. The pattern of the brain lesions showed more plaques than spongy areas. The plaques were similar to the "florid" plaques of kuru and those seen in sheep with scrapie, but more widespread.
3. The electroencephalogram (brainwave) pattern was not like that for sporadic or iatrogenic CJD.

Most interesting was the discovery that all 10 of the victims had two valines at position 129 of the PrP gene.

Dr. Will sent a letter to all neurologists on March 21, detailing very specific signs and symptoms of the newly described disorder (see Box 4.1). The letter urged physicians to discard any instruments used to perform a brain biopsy or autopsy.

Although *The Lancet* report stressed that the evidence was not yet sufficient to establish a definitive link to BSE, panic set in anyway and beef prices plunged. Several farmers committed suicide as they watched their livelihoods vanish. Typically crowded fast food burger restaurants began importing beef and posting signs that said so. The British government argued with the European Union over how many cattle to destroy. And everyone awaited predictions of how many people would fall ill—a terrifying prospect, considering that more than 100,000 cows had died. At a November 26 meeting of the Royal Statistical Society in London, president Adrian Smith told the press that it would be "zero to a million" human cases, a pithy way of saying he had no idea. Somehow this translated into the newspapers and tabloids as "hundreds."

Evidence of a link between BSE and nvCJD mounted. The human outbreak was eerily like BSE, with the 10 young people suddenly and simultaneously becoming ill between

BOX 4.1

NEW VARIANT CREUTZFELDT-JAKOB DISEASE

The symptoms of new variant Creutzfeldt-Jakob disease are as follows:

- *Early age of onset (average 27.6 years, age range 18–41 years).*
- *Prolonged illness (average 13.1 months, range 7.5–24.0 months).*
- *Early neurologic signs and symptoms include depression, anxiety, behavioral changes, and pain or numbness when limbs or face are touched.*
- *Loss of coordination.*
- *Memory loss, inability to move.*
- *Twitching and other uncontrollable movements. The electroencephalographic (EEG) pattern of sporadic or iatrogenic CJD is absent.*
- *Extensive plaques throughout the brain, as well as spongiform change, neuronal loss, and astrocyte gliosis (overgrowth of supportive star-shaped cells).*

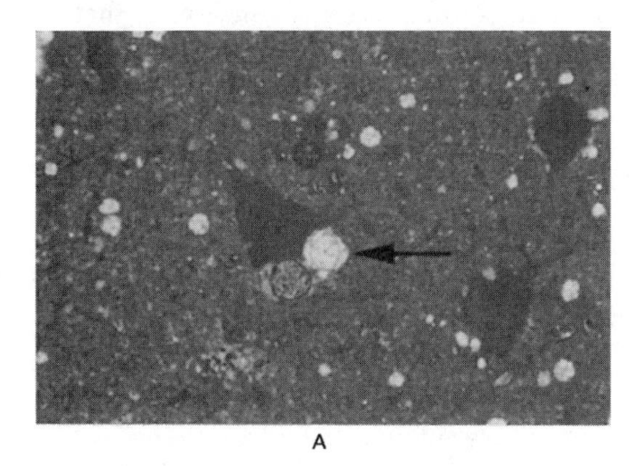

A

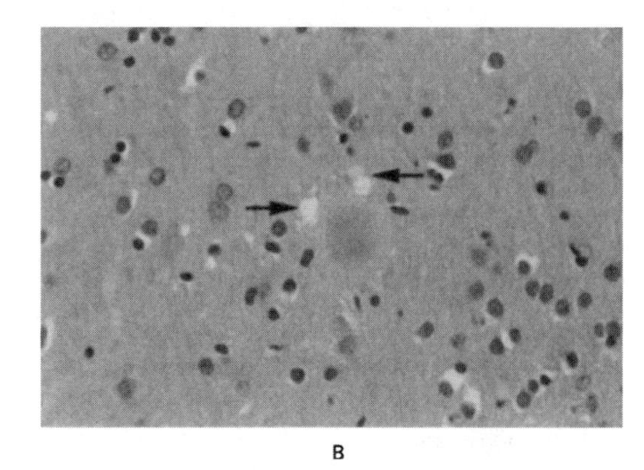

B

FIGURE 4.3

In contrast to the brains of people with sporadic CJD (A), those from people with new variant CJD have characteristic florid plaques (B).
Reproduced with permission from Johnson T and Gibbs J Jr. Creutzfeldt-Jakob disease and related transmissible spongiform encephalopathies. NEJM *1998;339:1994.*

February 1994 and October 1995 just as the cows had begun to suddenly sicken a decade earlier. Biochemical data also linked the two diseases.

At the Imperial College School of Medicine in London, John Collinge and coworkers examined PrP in the brains of the nvCJD victims (Figure 4.3). The prions were unlike those of other types of CJD, but very similar to those of BSE after passaging them through mice and other mammals. The researchers used a technique called Western blotting to examine the protein's shape, particularly the pattern of sugar molecules attached to it, which constitutes a particular glycoform. The glycoform of nvCJD looked exactly like that of BSE, but not like the agent that causes sporadic and iatrogenic CJD. This finding supported the conclusion that the prion "strain type"—defined by its effect on the host—was a function of the prion's conformation and the pattern of sugars linked to it.

1997

Collinge's group and the Edinburgh team came up with even more convincing evidence of a BSE-nvCJD link. The cover of the October 2 issue of *Nature* proclaimed there was "A common agent for BSE and vCJD." (The name apparently segued from "new variant" to "variant" as the disease became not so new.) Their conclusion: The biological and transmission characteristics of the new prion made it "consistent with" its being the human counterpart of the prion that causes BSE.

Collinge's technique was to use transgenic mice that manufacture human PrP that has a double dose of valine at position 129. These transgenic mice were therefore a model of sorts for a human susceptible to prion infection. The researchers knew that the transgenic mice easily acquire "old" CJD, but nontransgenic mice don't. However, the transgenic mice took a long time to develop variant CJD, and when they did they had a new, strange symptom—they walked backwards. When the researchers exposed transgenic mice to the BSE prion from cows, the same thing happened—the mice walked backwards. The incubation times and pattern of brain lesions also corresponded between vCJD and BSE. The message was chilling: If the prion that causes BSE could affect and infect mice that had human PrP genes, it could probably infect humans too.

1998

By 1998, 27 cases of vCJD had been reported, although this was surely an underestimate. In June, Tony Barrett, a 45-year-old coast guard from southwestern England, died of vCJD and provided researchers with a new type of tool—his appendix, removed and saved before symptoms arose. The appendectomy was performed in September 1995, and Barrett first noticed that his hands and face were numb in May 1996. Lymph nodes in the preserved appendix had abnormal PrP. By this time Collinge had developed a way to detect scrapie PrP in tonsils. He wasn't surprised to find prions in the lymphoid tissue because he had detected them in tonsils of sheep with scrapie. Sheep tonsils yielded scrapie prions when the animals had progressed only a third of the way through the several-year incubation period.

1999

The possibility that vCJD could occur outside the United Kingdom arose. Douglas McEwen, a 30-year-old deer hunter living in Utah, lay dying as 1999 began. He had begun to lose the ability to spell familiar words and to do simple mathematics the previous June. The usual blood tests and medical scans were negative, but a brain biopsy performed in November revealed vCJD. By this time, McEwen could no longer recognize his wife or daughters.

Several clues pointed to vCJD. McEwen declined swiftly, was too young to have sporadic CJD, and had never had a transplant or been touched by prion-tainted surgical instruments. Nor was he a cannibal. But he also had never eaten British beef. He had hunted deer and elk, however, animals that develop a neurologic disorder called chronic wasting disease. Like the mad cows, affected deer and elk drool, stagger, lose weight, and die. Although chronic wasting disease has been detected in the wild only among deer and elk in southern Wyoming and northeastern Colorado, McEwen could have encountered an animal from there or from other affected wild populations not yet recognized. Plus, the disease occurs in captive deer and elk populations in Oklahoma, Nebraska, South Dakota, and Saskatchewan. These cases were traced to the 1960s, when animals from wild Colorado populations were sent to zoos and game farms while the disease was silently incubating.

FATAL INSOMNIA

As if scrapie, kuru, and the CJDs weren't odd enough, another prion disease was joining the group as BSE was taking its toll on the cows of England. In 1986, Pierluigi Gambetti (today at Case Western Reserve University) and colleagues reported in the *New England Journal of Medicine* a case of "fatal familial insomnia" (FFI). It was as horrible as it sounds, reminiscent of episode 91 of *Star Trek: The Next Generation.* In "Night Terrors," which aired March 16, 1991 (Stardate 19144631.2), the ship's wandering into a rift in space deprived crew members of dream sleep, causing terrifying hallucinations and extreme paranoia. But unlike Commander Riker and colleagues, people with FFI do not sleep at all. Nor do they recover at the end of the episode.

The 1986 paper reported on a 53-year-old Italian man. His illness began with inability to fall asleep at night, which progressed rapidly to nearly total insomnia. His autonomic nervous system was malfunctioning too, producing fever, excess sweating and tearing, diminished reflexes and sensation of pain, and skin blotches. His coordination deteriorated and, from the lack of sleep, he became emotionally unstable, began to hallucinate, and became delusional. Then he developed tremors, couldn't speak normally, went into convulsions, and, nearing nine months since the onset of symptoms, fell into a coma and died. Like the victims of kuru, he remained rational and aware for most of the course of the illness.

The pattern of damage to the man's brain was quite distinctive. Degenerated neurons and overgrowth of astrocytes appeared only in a specific part of the thalamus, a relay station of sorts in the brain. The result was paralysis of areas that control sleep, certain involuntary functions, and circadian rhythms. The brain differed from the brain of a vCJD victim in that it lacked spongy areas and the affected area was very localized.

Other family members, including two sisters, suffered the same strange collection of symptoms, and so the fatal insomnia was dubbed "familial." Scanning the literature, Gambetti uncovered a paper published a few months earlier describing "familial myoclonic dementia masquerading as CJD," which was probably the same disorder. Whatever the name, this bizarre disorder is extremely rare. A few more cases found in 1990 in France and Belgium turned out to be relatives of the first family. Only 24 affected extended families have been identified to date. The disease begins anywhere from age 25 to 61 (average 48 years), and the duration is from 7 to 35 months (average 18 months).

The brain damage pattern and some of the symptoms suggested that FFI was a prion disease. Molecular evidence confirmed this. In 1992, researchers found that antibodies to scrapie PrP bind to brain material from FFI patients. But the prions associated with the insomnia were unique. Exposing them to protease digestion yielded a different-sized "protease-resistant core" than other known human abnormal prions. The core is the part of the protein that remains after exposure to protease.

Because the fatal insomnia was inherited, looking at the PrP gene sequence could provide clues to what was happening at the protein level. Perhaps this information might

TABLE 4.2

A Comparison of Fatal Familial Insomnia and Variant Creutzfeldt-Jakob Disease

Disease	Amino Acid 178	Amino Acid 129	Protease-Resistant Core (kd)	Brain Damage
FFI	Asparagine	Methionine only	19	Thalamus, no spongy areas
vCJD	Asparagine	Valine only	21	Throughout

TABLE 4.3

Progression of Fatal Sporadic Insomnia

Time from Onset	Symptoms
4 months	Sleep 1 hour per night
6 months	Excess tearing, difficulty walking
10 months	Inability to cough, difficulty speaking, loss of short-term memory, repeated statements
12 months	Wheelchair-bound due to poor coordination, extreme difficulty speaking, inability to separate dreams from reality
16 months	Jerky movements, severe hallucinations and delusions, death from aspiration pneumonia

apply to sporadic prion disorders too. The 53-year-old man with FFI had a mutation at codon 178 in the PrP gene, which replaced the amino acid aspartic acid with asparagine at this position in the protein. Researchers were at first perplexed, because this is the same amino acid change that John Collinge's group had seen in growth hormone–induced CJD. How could the same alteration in the amino acid sequence of PrP cause two different disorders—FFI and CJD? The answer came from looking elsewhere in the protein, to the 129th amino acid. A change here had been seen before as well—this is the site in the protein where victims of vCJD have two valines.

People with vCJD or FFI share the aspartic-acid-to-asparagine change at amino acid 178. If they have two methionines at position 129, they develop FFI. But if they have two valines at position 129, they develop vCJD. In addition, the protease-resistant core for vCJD is larger than that for FFI. They are two distinct diseases, but are both caused by compound mutations in the same gene (Table 4.2).

In the pre-BSE days of 1986, identification of fatal insomnia didn't capture much media attention, and its rarity perhaps labeled it an oddity in the medical community. But when another report of fatal insomnia appeared in the May 27, 1999, issue of the *New England Journal of Medicine,* after the BSE scare, more people took note. James Mastrianni in Stanley Prusiner's lab reported on a 44-year-old man from California who, like the man described 13 years earlier, suddenly experienced an inability to fall asleep. After a week, he consulted his doctor, who told him it was "all in his head." It was, but not in the way that the doctor thought. Four months later, the insomnia had worsened to one hour of sleep a night, and the course was downhill from there (Table 4.3).

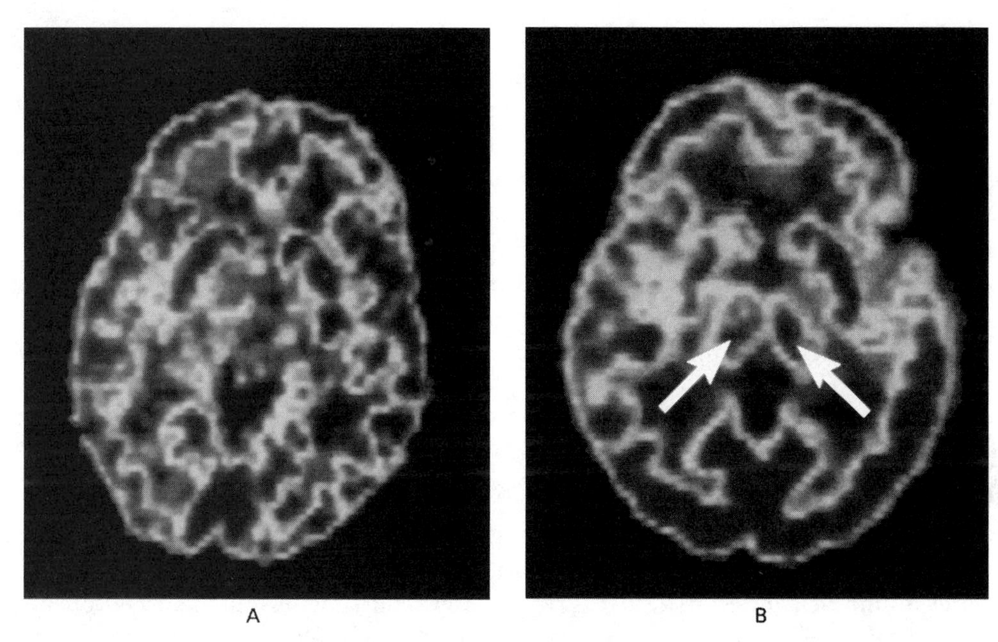

FIGURE 4.4

A PET scan of the brain of a man who died from fatal sporadic insomnia (A) showed lack of metabolic activity in the thalamus, the region indicated by arrows in the brain from a healthy individual of the same age depicted in (B).
Reproduced with permission from Mastrianni JA, Nixon R, Layzer R, et al. Prion protein conformation in a patient with sporadic fatal insomnia. NEJM *1999;340:1630.*

Tests finally conducted a year after the man first reported his insomnia, including a magnetic resonance imaging (MRI) scan and a biopsy of the cerebellum, were normal. A positron emission tomography (PET) scan, however, revealed greatly diminished uptake of radioactively labeled glucose in the thalamus, the same pattern of damage seen in the 1986 case of fatal familial insomnia. The correspondence between the two cases continued with the autopsy. The 44-year-old man's brain, when compared with the brain of a similarly aged man who had not died of a neurologic disorder, showed the same areas of the thalamus devoid of neurons and overrun with astrocytes (Figure 4.4). Antibody staining revealed scrapie PrP here, but at concentrations less than in a vCJD brain.

What was most significant about the 1999 case was that it was *not* inherited. The man had no known affected relatives, and his PrP gene was normal. "We found a disease that is indistinguishable from the genetic disorder, but lacks the disease gene," says Mastrianni. Yet the prions that were eating away at the man's thalamus were definitely not normal. Although they yielded the same-sized protease-resistant core as the prions from the man with FFI, the sugars were attached in a different pattern. Prusiner and colleagues named this new guise of the *Star Trek* torture "fatal sporadic insomnia," and the conclusion was compelling, validating years of conjecture and research: If this man's PrP gene was normal, yet he had the same symptoms as the man with the familial form, then it is the conformation of the protein that causes illness, and not the amino acid sequence. "Even in the familial form, it's the prion strain, the precise conformation of the misfolded protein, and not the variation in the gene that ultimately determines the consequences of the disease," concludes Mastrianni.

How the man developed the sporadic fatal insomnia is a mystery. Perhaps he harbored a mutation in a protein, called a chaperone, that normally guides the folding of another protein, in this case PrP. Or, a somatic mutation may have occurred in a neuron or astrocyte in the thalamus, rendering only a portion of the brain able to produce scrapie PrP. Alternatively, perhaps the prion protein rarely and fleetingly assumes an abnormal conformation that can become stabilized. The prion's unusual shape is not only stabilized, but infectious, serving as a template to convert normal PrP.

As fatal sporadic insomnia was being probed, elegant experiments elsewhere at UCSF were catching the responsible shape-shifters in action.

VISUALIZING PRIONS

Genetics provides indirect clues regarding how normal PrP becomes the abnormal scrapie form. The fact that many victims of prion diseases have only valine or only methionine at amino acid position 129 in PrP suggests that sameness at this site somehow stabilizes the evil twin conformation. Most people are heterozygous at this position, which means that some copies of PrP bear valine, and some methionine.

Another clue to the chemical basis of prion diseases comes from the inherited Gerstmann-Straussler-Scheinker disease, which causes lack of coordination, and sometimes dementia, over an 8- to 10-year period until death. In affected individuals, the PrP gene has extra copies of an 8-peptide repeat, called an *octapeptide*. The normal gene has 5 octapeptides, located between amino acids 51 and 91 in the sequence. Individuals with the disease may have up to 14 such repeats in the PrP gene. This finding was disturbingly familiar to geneticists. Similar repeats that destabilize other proteins that gum up the brain have been well studied since the mid-1980s in the "triplet repeat" disorders. First described in myotonic dystrophy and fragile X syndrome, this class of mutation came to be epitomized by Huntington disease, in which uncontrollable movements and personality changes begin at around age 38. Affected people have from 35 to 121 copies in a row of the amino acid glutamine inserted into a protein called huntingtin. The normal number ranges from 6 to 37. Although the triplet repeat disorders are not prion diseases, they seem to share with these conditions a fundamental glitch in the folding of a particular protein.

Glimpsing precisely how the prion protein contorts out of normalcy, and perhaps exposing an Achilles' heel, required an imaging technique. X-ray crystallography is the gold standard in revealing a protein's conformation, but obtaining prion crystals proved challenging, if not impossible. So instead, UCSF's Thomas James turned to nuclear magnetic resonance (NMR) imaging (Figure 4.5). "The protein was prepared in the lab of Stan Prusiner. We would certainly have determined the structure earlier, but for nearly a year almost all of the protein was consumed in a fruitless effort to get crystals. But with what we know about the protein structure now—that it is large with relatively little stable structure—it may be quite difficult ever to get crystals," James says.

NMR indeed captured the changeling part of the prion protein. The technique detects signals from carbon atoms; the signals differ depending on how many hydrogens are in the vicinity. Integrating the data confirmed the presence and location of two common protein "conformational motifs" (shapes) that Prusiner's lab had identified in earlier studies, regions of alpha helices and beta pleated sheets. Further work revealed that copper

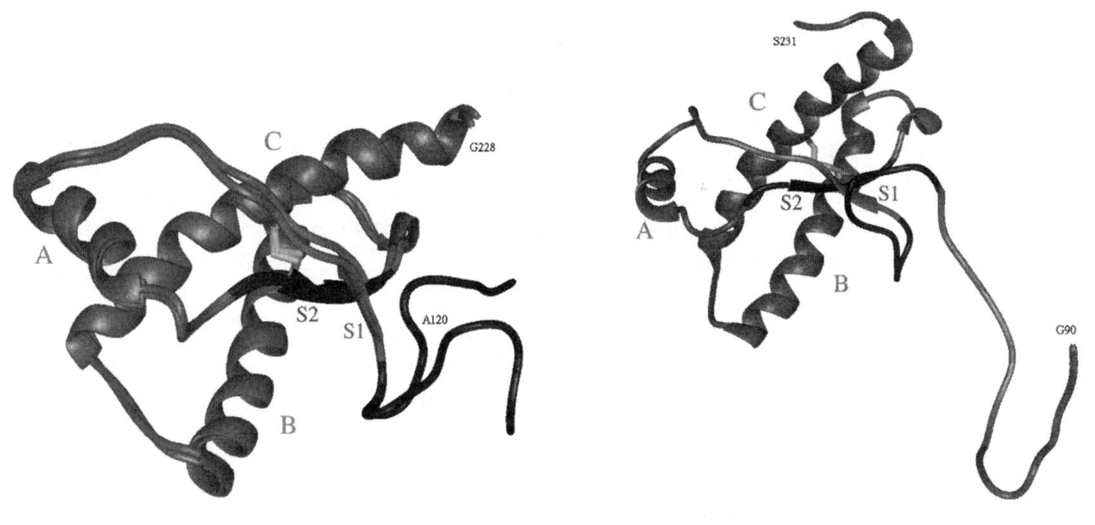

FIGURE 4.5

PrP can assume more than one stable conformation. Here are two normal "cellular" forms of the protein. In the scrapie form, one of the helices opens out and other changes occur that somehow make the particle infectious.
Courtesy of Thomas L. James, University of California, San Francisco.

atoms bound to the octapeptides fold the amino acid chain into the helices. The NMR data zoomed in on the structure.

Prion protein has a core domain and a backbone of three alpha helices and two short head-to-toe beta sheets. "Our structural studies show that while most of the core of the prion protein assumes a single, stable structure, an important part of the prion protein exhibits multiple structures, in at least two conformations," says James. Peptide fragments from the unstable region assume the normal cellular form of PrP under some solution conditions, but change into the infectious scrapie form, with some of their alpha helices opening to form beta sheets, under others.

The NMR structure also revealed how the aspartic-acid-to-asparagine switch at position 178 and the amino acids found at position 129 affect PrP. One end of the aspartic acid at position 178 attaches to a ring that is part of the amino acid tyrosine at position 128, next to the all-important 129. Substituting asparagine for aspartic acid at position 178 disrupts the bond with the tyrosine at 128, which produces a different overall shape to the molecule depending on whether position 129 is always methionine, always valine, or both methionine and valine. Amino acid 129 is indeed the linchpin, located in the region of the prion protein that can assume multiple shapes.

THE YEAST CONNECTION

Biologists often study a complex phenomenon using a simple system that can serve as a model. In this way, bacteria helped unravel the general workings of genes, simple animals such as roundworms and sea urchins revealed much about development, and mice are providing clues to many human diseases. The yeast *Saccharomyces cerevisiae* may be a compelling model for the prion disorders that seem to be unique to mammals.

Of course, yeast, being single-celled organisms, hardly develop the Swiss cheese brains that are the hallmark of the spongiform encephalopathies. But they do have at least two genes that encode proteins that assume different shapes without a change in the amino acid sequence—just like PrP. Also like classic prions, once a yeast protein assumes its abnormal form, it spreads the change. Yeast prions do not bear any chemical similarity to mammalian prions, except for containing repeated sequences. The "traits" that they provide are biochemical in nature: One abnormal form disrupts the termination of protein synthesis, and the other interferes with a cell's ability to break down nitrogen-containing compounds.

The yeast prions have been known for several years, but it is only recently that they have been subject to genetic engineering experiments that fully support Prusiner's still controversial ideas. Based on studies on yeast, in 1994 Reed Wickner at the National Institute of Diabetes and Digestive and Kidney Diseases outlined three requirements for prions:

1. They form spontaneously.
2. Overproduction of the normal form of the protein increases the chances of an abnormal form arising.
3. A mutation in a prion gene that prevents the normal protein from forming also prevents the abnormal protein from forming.

A consequence of these "rules" is that the altered characteristic appears the same whether it derives from a mutation in the gene or from spontaneous formation of the prion form. This is true for yeast and for people with inherited versus acquired prion diseases.

Susan Lindquist, a Howard Hughes Medical Institute molecular biologist at the University of Chicago, took a closer look at just what a yeast prion gene can do by removing the part of it that induces infectivity and hooking it up to a totally unrelated gene that controls hormone secretion in rats. She essentially created a new type of prion, a little like linking a new train to an engine that usually carries something else. In another experiment, she altered the gene to contain extra repeats and found that abnormal prions were more than four times as likely to form. In addition, the abnormal form of a yeast prion protein misfolds in a way that causes others to stick to it, creating fibers of amyloid not unlike those seen in some CJD brains, as well as in the brains of people with Alzheimer's disease, advanced diabetes, and atherosclerosis. The idea that some sort of infection, or spontaneous propagation of an abnormal protein form, may underlie some of these disorders is not as farfetched as it was regarded just a short time ago—illustrating, once again, that a compelling experiment can overturn a long-cherished idea in science.

WHAT'S NEXT?

Despite the piles of accumulating evidence that proteins can indeed cause infectious disease, which were bolstered by the yeast findings, the definitive experiment—coaxing normal PrP to assume the scrapie form in a test tube and then demonstrating that it causes illness in an animal—has yet to be accomplished as of this writing. "Several labs are trying to carry that out, but getting it to happen in an unambiguous way may be difficult. It is a very unusual molecular transformation that could be hard to reproduce in the test tube," says David Harris, an associate professor of cell biology and physiology at the Washington University School of Medicine in St. Louis. Difficulties include the long incubation time and the effects of other molecules that may be necessary to convert normal to abnormal protein. Harris has taken mutant PrP and watched it convert normal PrP in cultured cells and in transgenic mice, definite steps in the direction of those long-sought definitive experiments.

We still do not know exactly what normal PrP does. Mice bred to lack it have only minor abnormalities, such as disrupted circadian rhythms, suggesting that it isn't vital. Therefore, sheep and cows bred to lack the gene might be incapable of developing scrapie or BSE, but might otherwise be fairly healthy. However, neurons growing in culture die prematurely when they lack normal PrP. Susan Lindquist suggests that prions are the manifestation of a sort of flexibility built into protein structure—in a sense, two for the price of one. She calls prions "an ancient mechanism of inheritance, but a newly appreciated one." But under which environmental circumstances an abnormal prion form is an advantage (and presumably this is the case because it has been retained through evolution) remains to be seen.

In a historical sense, humans have been lucky as far as prion diseases go. The growth hormone route to CJD was a near-disaster. Variant CJD has claimed only a few dozen lives that we know of so far, compared with the 100,000-plus cows stricken with BSE. Sporadic CJD afflicts one in a million, and iatrogenic CJD can be avoided by steering clear of medical instruments tainted with infectious brain matter from past patients. Still, kuru's dramatic skewing of the sex ratio in some New Guinea villages indicates that a prion disease always has the potential to decimate a human population.

Of course, we don't know if the prion diseases are really as rare as they appear to be. Pierluigi Gambetti suggests that fatal sporadic insomnia may actually account for many cases of dementia traced to the thalamus. And how many cases of CJD go undiagnosed, as they did in 1997 when the woman with lung cancer donated her corneas, with no one noting her dementia at the time of death? Other cases of what are really prion diseases may be misdiagnosed as early Alzheimer's disease.

Will we be able to prevent and treat prion diseases in the future? For a start, surgical patients and potential organ donors will be much more carefully checked for signs of prion illness. Diagnosis of prion diseases will also occur earlier in the course of the illness, because physicians finally know what to look for. And the finding that certain compounds can stabilize the nonpathogenic form of PrP suggests that a drug treatment may be feasible.

Future epidemiologic research will seek to identify cases of prion infection that have not yet caused symptoms. The British government is planning a huge study to test stored appendix and tonsil samples to determine whether vCJD infection was really confined to a few dozen people. And in the United States, researchers are keeping an eye on deer and elk hunters who develop suspicious symptoms.

Finally, learning how prion diseases develop may explain how other, more common conditions that clog the brain arise. "In the amyloid diseases, such as Alzheimer's, Parkinson's, ALS, Huntington's, as well as the prion diseases, a normal protein partially unfolds, then refolds into large plaques and fibrils that are held together via intermolecular beta sheets. This may be a common theme," says James. In fact, Gajdusek and Prusiner's funding from NIH has been largely based on the similarities between the prion diseases and more common disorders.

"NIH acknowledged the uniqueness of the prion protein, but also recognized the broader implications. There are a number of neurodegenerative conditions, at least some subset of which are due to a general underlying mechanism. We think these disorders represent aggregations of parts of proteins or entire proteins, that aggregation somehow leads the cell, particularly a neuron, to degenerate. How it does so—that's the big black box. A handle to the causes of these diseases is somehow related to a structural change that converts a normal protein to a scrapie-like protein," says D. Stephen Snyder, director for the Etiology of Alzheimer's Disease program at the National Institute on Aging. All of these conditions—the prion disorders, ALS, Huntington and Parkinson's diseases—are collectively termed "protein conformational diseases."

The discovery of prions offers a compelling message about the process of science: We should not ignore clues that do not fit an established way of thinking, for they may be telling us something new. Says Snyder, "The story of prions illustrates that we don't know where basic science will lead. Kuru was once looked at as a unique, abstruse, and foreign disease. Now 20 years down the road, it is applicable to diseases like Alzheimer's. But that's how science works—it's unpredictable."

REFERENCES

Aguzzi, Adriano. "Protein conformation dictates prion strain." *Nature Medicine* 4:1125–1126 (1998).

Anderson, R. M., et al. "Transmission dynamics and epidemiology of BSE in British cattle." *Nature* 382:779–788 (1996).

Balter, Michael. "On the hunt for a wolf in sheep's clothing." *Science* 287:1906–1908 (2000).

Brown, Paul. "The risk of bovine spongiform encephalopathy ('mad cow disease') to human health." *Journal of the American Medical Association* 278:1008–1011 (1997).

Collee, J. Gerald, and Ray Bradley. "BSE: A decade on." *The Lancet* 349:636–641 (1997).

Collinge, James, et al. "Genetic predisposition to iatrogenic Creutzfeldt-Jakob disease." *The Lancet* 337:1441–1442 (1991).

Gajdusek, D. Carleton. "Unconventional viruses and the origin and disappearance of kuru." *Science* 197:943–960 (1977).

Gambetti, Pierluigi, and Piero Parchi. "Insomnia and prion diseases: Sporadic and familial." *New England Journal of Medicine* 340:1675–1677 (1999).

Johnson, Richard T., and Clarence J. Gibbs Jr. "Creutzfeldt-Jakob disease and related transmissible spongiform encephalopathies." *New England Journal of Medicine* 339:1994–2004 (1998).

Lewis, Ricki. "Prions' changeability: Nuclear magnetic resonance shows more pieces of the puzzle." *Scientist,* 9 May 1999, 1.

Li, L., and S. Lindquist. "Creating a protein-based element of inheritance." *Science* 287:661–664 (2000).

Mastrianni, James A., et al. "Prion protein conformation in a patient with sporadic fatal insomnia." *New England Journal of Medicine* 340:1630–1638 (1999).

Prusiner, Stanley. "Molecular biology of prion diseases." *Science* 252:1515–1521 (1991).

Prusiner, Stanley. "The prion diseases." *Scientific American* 277:48–57 (1995).

Sigurdsson, Björn. "Rida, a chronic encephalitis of sheep." *British Veterinary Journal* 110:341–355 (1954).

Soto, Claudio. "Alzheimer's and prion disease as disorders of protein conformation; implications for the design of novel therapeutic approaches." *Journal of Molecular Medicine* 77:412–418 (1999).

Will, R. G., et al. "A new variant of Creutzfeldt-Jakob disease in the UK." *The Lancet* 347:921–925 (1996).

THE TALE OF TELOMERES

Aging is an inevitable part of life. Although aging is a life-long affair, we usually first notice the gradual breaking down of structures and slowing down of functions in middle age. Our blood vessels narrow, our hair thins, our skin wrinkles, and our immunity wanes. These signs of aging all too obviously strike at the whole-body level, yet aging happens at a microscopic level too.

A complex animal such as a human is an amalgam of many cells, of many types. These cells have a built-in "sense" of their aging, a molecular counter of sorts that sets how many more times they can divide before they either settle into an extended existence as they are or start to show distinctive signs of senescence, the aging process. This molecular counting device is seemingly simple: The ends of chromosomes, called telomeres, *shorten each time a cell divides, until they reach a point that signals "stop" (Figure 5.1).*

In science, though, the seemingly simple is rarely so—it just appears that way in retrospect, after we learn how something works. The road to understanding telomere function began with observations of the chromosomes of corn and flies, then jumped to human fetal cells inexorably aging in a laboratory dish, continued with clever experiments on one-celled pond dwellers, and proceeds today in studies of a variety of organisms. Along the way, telomere biology, once an offshoot of cytology, has evolved into a medical science, shedding light on the origin and progression of cancer and degenerative diseases, as well as the very nature of the aging process.

DISCOVERING STICKY ENDS

Cell biologists consider the first decades of the twentieth century as the golden age of cytology, a time when chromosomes began to reveal their mysteries. Mendel's basic laws of inheritance had been rediscovered at the turn of that century, and the term *gene* coined

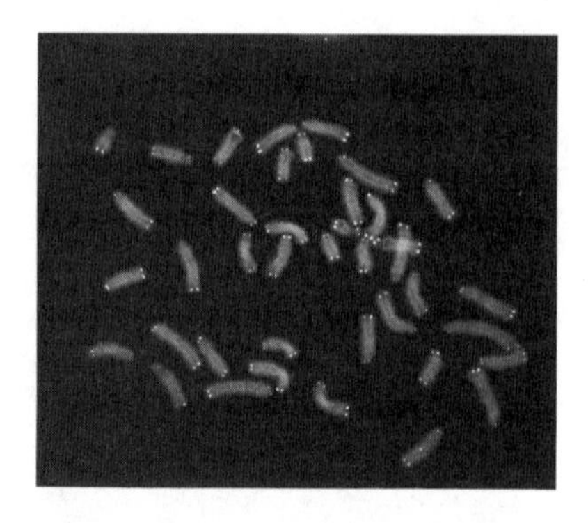

FIGURE 5.1

Telomeres.
Courtesy of Jerry Shay.

in 1908, but it would be another 40 to 50 years until researchers fathomed the molecular structure of a gene. The early cytologists, however, didn't need to know about DNA to prepare and observe chromosomes. Two such visionaries were Barbara McClintock and Herman J. Muller.

In 1938, McClintock was working with maize at the University of Missouri. She noticed that when the ends were cut off a chromosome on the brink of cell division, it would either fuse with another detipped chromosome, fuse with its newly generated sister, or close up and form a ring. Either way, the resulting divided chromosomes could not be equally distributed into daughter cells, instead forming bridgelike structures that would break anew. This cycle of breakage-fusion-bridge continued. McClintock's conclusion was that chromosome stability requires chromosome tips.

At about the same time, Herman J. Muller was working with chromosomes of the fruit fly *Drosophila melanogaster* at the Institute of Medical Genetics in Edinburgh. Like McClintock, Muller implicated loss of chromosome tips in causing all sorts of abnormalities. In a talk at the Marine Biological Laboratory at Woods Hole, Massachusetts, on September 1, 1938, Muller coined the name *telomere* from the Greek word *telos,* meaning "end," and *meros,* meaning "part":

> [T]he terminal gene must have a special function, that of sealing the end of the chromosome, so to speak, and that for some reason a chromosome cannot persist indefinitely without having its ends thus "sealed." This gene may accordingly be distinguished by a special term, the "telomere."

Thus was born telomere biology, although no one could then imagine exactly what a telomere was or how it differed from the rest of the chromosome. Muller suggested that a telomere's distinctiveness stemmed from the fact that unlike other genes, it had only one neighboring gene, like a person sitting in an aisle seat in an auditorium. But probing telomere composition was beyond the limits of the technology of the time.

Jump ahead to 1953. Years of elegant experiments had by this time overturned the long-held notion that the genetic material is protein, instead identifying it as DNA. In this year, Watson and Crick described the three-dimensional structure of the molecule of heredity, founding the field of molecular biology. A new crew of researchers, some emigrating from the physical sciences, delved into the double helix, teasing apart how the huge DNA molecule both replicates and controls transmission of inherited traits.

Meanwhile, telomeres remained only fuzzily defined, if at all. Textbooks at the time stated only that their absence led to "stickiness." In *The Chromosomes,* M. J. D. White, a professor of zoology at the University of Melbourne, wrote, "Some doubt still exists as to whether telomeres are actually visible in some instances as discrete bodies at the ends of the chromosomes, but the term is justified by the special behavior of the chromosome ends, even if no individualized structure can be seen." Even vaguer, Walter V. Brown and Eldridge M. Bertke wrote in their *Textbook of Cytology* in 1969 that "there is a theoretical uniqueness at the tip of each arm called the telomere." They noted that telomeres consist of dark-staining material, or heterochromatin, "whatever that really means."

And so as the 1970s dawned, telomere structure was still descriptive (dark-staining) and telomere function still an ill-defined default option (nonstickiness). But unrelated clues to the true nature of telomeres were already unfolding in a laboratory in Philadelphia as well as in the flurry of activity to learn how DNA replicates.

THE HAYFLICK LIMIT

If biologists were unsure of the nature of the material capping chromosomes, they were pretty certain, although incorrectly it would turn out, of something else—that normal cells growing in culture can divide forever. Biologists had been nurturing cells from various organisms in glassware since the early 1900s. "Over the subsequent 60 years, a belief grew, which then became dogma, that all cells in culture have a capacity for immortality. If cells in culture do die—and cells in culture do die a lot—it was because we were ignorant of the proper conditions needed so that the cells would divide constantly. And because people thought normal cells were immortal, aging was thought to have nothing to do with intracellular events," recalls Leonard Hayflick, today a professor of anatomy at the University of California at San Francisco. Instead, biologists thought that aging triggers came from outside, in the form of viruses, radiation, hormones, and other biochemicals in the milieu surrounding cells.

Senescence was widely thought to happen to bodies, not cells. Typical was the description of cell death in *The Science of Life*, a 1934 textbook by H. G. Wells, Julian S. Huxley, and G. P. Wells. (H. G. Wells became better known for his science fiction works, such as *The Time Machine*.) They wrote:

> [O]ur cells do not die because mortality is inherent in their internal structure. They die because they are parts of a very complicated system based on co-operation, and sooner or later one of the tissues lets the other down . . . . As a matter of fact, living matter is potentially immortal. If one keeps a culture from the tissue of a young animal and takes sub-cultures regularly, the race of cells can apparently go on growing and dividing indefinitely.

It turned out that cells dividing forever was not, in fact, a matter of fact. Leonard Hayflick found just the opposite, thanks to many experiments, the help of a talented associate, serendipity, and the insight to recognize the importance of what he was observing and to pursue it even in the face of persistent opposition—a recurring theme in scientific discovery. The situation was a little like Woese's observations of the unusual ribosomal RNAs of the archaea (see Chapter 3) and trying to explain repeated unexpected results as exceptions to a rule, rather than realizing that a different rule applied. The same sort of paradigm shift in thinking grew from the investigation of prions, the proteins with the unlikely property of infectiousness (see Chapter 4).

Leonard Hayflick's encounter with the challenge of having to disprove a popular idea in science began in the early 1960s, when he was working at the Wistar Institute in Philadelphia. He wasn't initially interested in cellular aging at all, but was one of many investigators searching for viral causes of cancer. His research plan was to take fluid from around cancer cells growing in culture and see what would happen if he transferred it to normal cells. The best human cells to work with came from embryos and fetuses because they are relatively free of viruses, which would add unwanted variables to the experiments. "Fortunately, I was in a position to pursue the idea. Today if I did that using federal research dollars to grow tissue from human embryos or fetuses, I would go to jail," says Hayflick, referring to the National Institutes of Health's constantly changing policy on the use of federal funds to derive cells from human embryos or fetuses, discussed in Chapter 6 in the context of stem cell research.

Ideally, several replicates of experiments transferring the cancer cell soup should have been started at the same time, but this wasn't possible due to the erratic supply of human fetal tissue. So Hayflick had to set up the experiments whenever he could. "In my incubator at any given time, I'd have 12 to 20 cultures going, each marked with a different start date," he recalls. But he quickly noted that "something unusual was going on" with the supposedly immortal normal human fetal cells in his incubator. "Despite the fact that I used the same technician, the same glassware, and the same media, the cells in culture the longest stopped dividing, while the young cultures luxuriated. That shouldn't happen. It intrigued me, so I began to look at what was going on." From that instant, he never went back to pursuing the viral cause of cancer—he was onto something much bigger.

What Hayflick was seeing in the time-staggered experiments was that fetal cells died after they had been transferred to new glassware a certain number of times. (The transfers were done when the cells had divided sufficiently to coat the inside surfaces of the container—they simply needed more room.) Time didn't matter; clearly it was the number of transfers that correlated to cell death. But why were the cells dying? After Hayflick's colleague, Paul Moorhead, examined the chromosomes to rule out some genetic fiasco as killing the cells, both men felt even more strongly that they had discovered a limit to the number of times normal cells could divide. Perhaps previous claims of the inherent immortality of cells in culture merely reflected repeated contamination by healthy, younger cells.

Hayflick and Moorhead discussed their findings and interpretations with other cell biologists. But established ideas die hard, and their colleagues insisted the two were at fault, that they had somehow botched the experiments, killing the cells. But that wasn't the case. "We repeated the work over and over, with the same results," Hayflick recalls.

Refusing to become discouraged, Hayflick and Moorhead devised an experiment to test the emerging hypothesis of a cell division limit by mixing cells from an older male fetus with cells from a younger female fetus. Time went by, and only female cells remained. Could a virus have come along and attacked only cells with a Y chromosome? More likely, Hayflick and Moorhead reasoned, was that the older cells, which just happened to be from a male, reached a cell division limit sooner than the cells from the female, and had died. But even publishing these results attracted little attention.

Hayflick and Moorhead were stymied. Researchers confronting the unexpected, especially when the unexpected is unaccepted, face a maddening predicament. Carl Woese eventually convinced most biologists that he had identified an entirely novel form of life with mountains of data, as did Stanley Prusiner and others for prions. Hayflick and Moorhead had plenty of experiments in mind, but they were uneasy. "Paul and I were both fresh out of postdoctoral positions, and to make a mistake while challenging a central dogma in the entire field of cell biology would have torpedoed any hope of a successful scientific career," Hayflick says.

So they came up with a very unconventional approach that both beautifully demonstrated the nature of scientific inquiry and eventually won over the skeptics. "We sent the luxuriating cultures—the young ones—to the grey eminences of the field. We would tell them, 'by May 20th to 30th, the cells are going to die.' When the phone started ringing between May 20th and 30th, we decided to publish. If our work went up in flames, we'd be in the company of the grey eminences," Hayflick recalls. He and Moorhead could hardly be criticized for being too sloppy to keep the cells alive when the leaders of the field, gifted with the very same cultures, couldn't do it either! Clearly, something was killing the cells according to a rigid schedule. But even this daring move could not budge some prominent cell biologists from the entrenched idea that normal cells in culture live forever, unless the culturists inadvertently kill them. Still, Hayflick decided it was time to publish.

"We sent the manuscript to the *Journal of Experimental Medicine,* a top journal at the time. It was rejected. And the editor, Francis Peyton Rous, who a few years later was awarded the Nobel prize for discovering the first cancer-causing virus, wrote, 'The largest fact to have come out from tissue culture in the last fifty years is that cells inherently capable of multiplying will do so indefinitely if supplied with the right milieu *in vitro.* Your paper is rejected.' *Experimental Cell Research,* however, published the paper unchanged. And it has been cited thousands of times since," says Hayflick.

That *Experimental Cell Research* paper launched the field of biogerontology and also introduced a cell line of fibroblasts (connective tissue cells) called WI-38 that would become the basis of vaccines against polio, rabies, rubella, and rubeola, saving millions of lives and making many millions of dollars for pharmaceutical companies. Not only did Hayflick not receive a cent in royalties for years, but he was arrested in 1975 for stealing government property—the WI-38 cells—which is a story unto itself. Finally, a lawsuit settled in his favor combined with changing times, notably in the form of the 1980 landmark decision of *Diamond v. Chakrabarty* allowing the patenting of cells, got Hayflick some funds, if not recognition. But the atmosphere was markedly different than it is today. "Intellectual property was simply given away by starry-eyed biologists because the taint of commercialism would ruin an academic career!" wrote Hayflick in 1998, with the hindsight of having seen academic biology become so intimately integrated with commercial enterprises that nowadays university departments are practically expected to spin off companies.

But the importance of the research described in the *Experimental Cell Research* paper transcended one investigator's premature encounter with economic reality. The work documented and quantified the aging of cells in culture, and getting those spectacular results was both tedious and time-consuming. In their standard protocol, Hayflick and Moorhead would take a bit of tissue from an embryo, add digestive enzymes to break it into cells, then put the cells into bottles containing a nutrient broth. Within a week, the cells had divided to form a one-cell-thick layer coating the bottle. To continue the culture, they would then distribute those cells into two new culture bottles. When those two bottle bottoms were covered, it meant that on average, each cell had divided (Figure 5.2).

Hayflick and Moorhead repeated the culture-transfer experiment with 25 strains of fibroblasts from human embryos. They noticed that by about the fortieth round of mitosis, the time between divisions lengthened, as if the cells were slowing down. Most fetal cells simply stopped by about the fiftieth division, a point that the researchers dubbed the "phase III phenomenon" but that eventually became known as the "Hayflick limit." Fifty cell divisions is more powerful than it might sound—if all the daughter cells lived, that would amount to 20 million metric tons of cells, Hayflick calculates.

The fidelity of this internal cell division pacesetter was astounding, as the researchers discovered when they added a step—freezing cells into a state of suspended animation. Once thawed, the cells went about their divisions as if time had stood still. "Freezing was a new technology then. If we reconstituted cells frozen at a particular doubling level, they demonstrated memory. If we froze a cell at 20 divisions for a year or two, it remembered the doubling level, and would take off where it left off when reconstituted. WI-38 cells were frozen for 38 years, and their memory has proved as good today as it was in 1962!" Hayflick says.

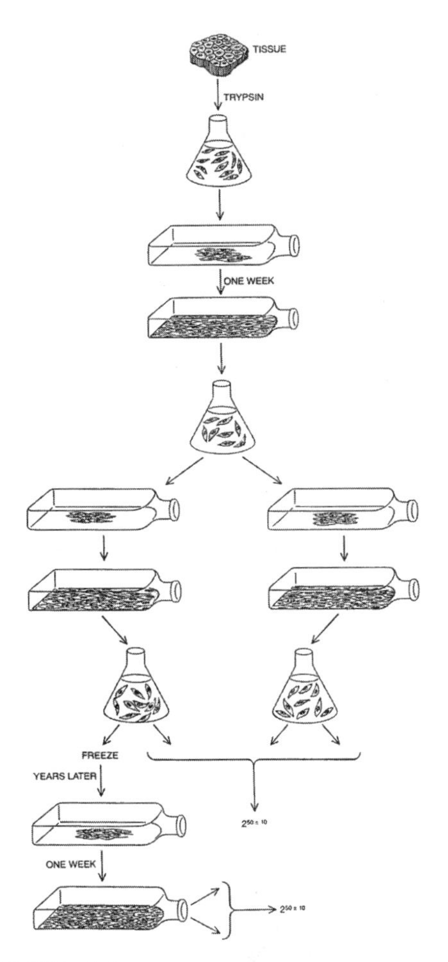

FIGURE 5.2

Hayflick and Moorhead demonstrated the finite lifetime of cells growing in laboratory culture with serial transfers. They began by teasing apart the cells of a tissue with a digestive enzyme, trypsin, and then transferred some of the cells to a bottle. After a week, the cells covered the bottom with a layer one cell thick. The researchers distributed these cells into two culture bottles and let them divide until the bottom was once again covered. The process was repeated, with each transfer representing another cell division. Normal cells can keep this up for about 50 divisions. If cells are deep-frozen, then thawed, they pick up where they left off, dividing the number of times that are left to reach 50.

Hayflick, L. "The cell biology of human aging," Scientific American *1980;242:58–65. Reproduced with permission from Alan D. Iselin.*

The *Experimental Cell Research* paper provided so much compelling evidence that the "grey eminences" could no longer ignore it. Hayflick and then others discovered by 1965 that the older the individual who provides cells, the fewer times those cells divide. And cells from people who suffer from rare rapid-aging conditions undergo fewer than their allotted 50 divisions, dying well ahead of the normal developmental schedule, as do their unfortunate hosts. Attention began to turn to discovering the nature of the counting mechanism. And as the 1960s drew to a close, researchers were flocking to the quest to identify what Hayflick would eventually call the cell's "replicometer." This is a more accurate term than "cell division clock," Hayflick maintains, because the mechanism counts DNA replications, not the passage of time.

In 1975, Hayflick and his doctoral student, Woodring Wright, implicated the nucleus. They took a nucleus from a cell that had divided 10 times and transferred it to a cell that had divided 30 times but whose nucleus had been removed. The recipient cell divided 40 more times, in tune with its new genetic headquarters. Although Hayflick didn't directly connect this control center in the nucleus to telomeres, he did note that as cells neared death, their chromosomes tended to break and fuse in a characteristic way, with pieces of different chromosomes annealing to form "dicentrics" (meaning "two centers"). Dicentrics are the only types of abnormal chromosomes associated with cellular aging. What Hayflick was observing was Barbara McClintock's "sticky ends," but with a new connection to aging.

At this point, telomere biology consisted of just two unlinked observations: Chromosomes without ends become sticky and unstable, and normal cells divide a set number of times. It would take discoveries and ideas from two other areas to complete the picture—musings into how DNA replicates, and a peculiarity of the chromosomes of certain protozoans.

THE END REPLICATION PROBLEM

In science, a major discovery rarely leads to closure, but instead triggers an avalanche of further questions. So it was for Watson and Crick's 1953 model of DNA structure. How did the immense molecule copy itself, a feat necessary to use the information held in its sequences of adenine (A), thymine (T), cytosine (C), and guanine (G) nucleotide bases? The solution would ultimately lead to the telomere.

Watson and Crick wrote and spoke about how the structure of DNA contained an embedded mechanism to replicate itself. Specifically, the two sides of the double helix would separate, and each side would pull in its complement, A with T and G with C (Figure 5.3).This mechanism was called "semiconservative" because each new DNA molecule would retain half of the previous generation's molecule. Semiconservative replication might make sense in theory,

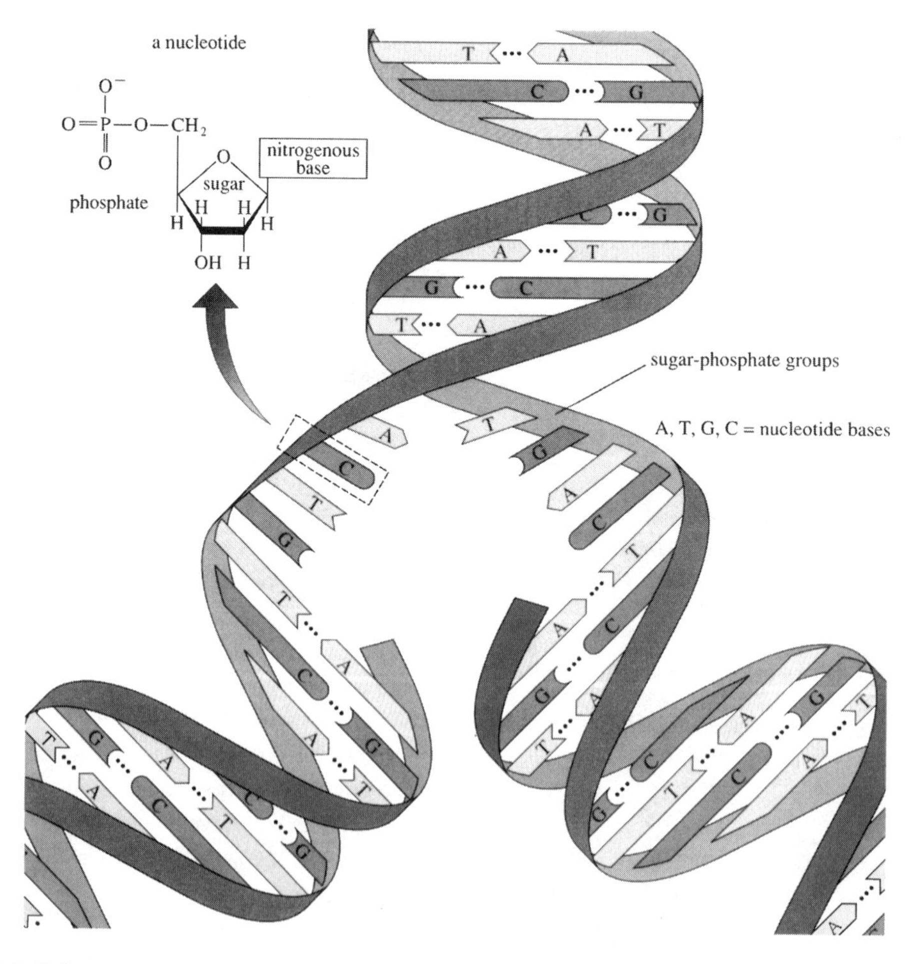

FIGURE 5.3

DNA replication is semiconservative; that is, each parental strand pulls in free nucleotides to build a daughter strand. The two sides of a DNA double helix are antiparallel, which means that they run in opposite orientations. This fact is important for the role of telomeres in DNA replication.

Lewis, R. Life, 3rd ed. Boston: McGraw-Hill, 1998:53. Reproduced with permission of The McGraw-Hill Companies.

but it was difficult to envision how it would actually happen, pointed out Max Delbruck of the California Institute of Technology. He wrote to Watson concerning "the replication problem," insisting that the double helix was too long and twisted to part as it replicated. Did the molecule contort temporarily to replicate, or routinely break and rejoin to relieve the tension? Delbruck suggested an experiment: Add a radioactive label to newly replicating DNA in an easy-to-study organism. The pattern of radioactive building blocks incorporated into new DNA would reveal the mechanism of replication. Predictions were possible.

If DNA replication were semiconservative, then a newly synthesized DNA double helix would be half labeled (the new strand) as well as half unlabeled (the parental strand). But if the molecule breaks and rejoins, each new double helix would be a mix of labeled and unlabeled parts, reflecting a "dispersive" mechanism. Or maybe a double helix just somehow doubles itself, using entirely new DNA, an approach called conservative. In 1957, a variation on Delbruck's idea yielded spectacularly clear results. Cal Tech's Matthew Meselson and Franklin Stahl labeled replicating DNA with a heavy isotope of nitrogen, and measured density rather than radioactivity. The second generation of DNA was a hybrid double helix, with one strand of normal density and the other heavy, demonstrating semiconservative replication.

But knowing that DNA replication is semiconservative didn't explain how an entwined double helix 2 meters long could methodically unravel to duplicate itself. The discovery of unwinding proteins ultimately explained that obstacle. But another barrier in understanding how DNA replicates was that the double helix is directional. That is, the sugar-phosphate backbone to which the base pairs attach runs in opposite directions on either side of the molecule, like a head-to-tail configuration. This orientation is called antiparallelism. The problem was that the enzyme—a DNA polymerase—that creates a new DNA strand can only work in one direction. How, then, could both sides of the double helix bring in new bases at the same time? In the early 1970s, further labeling experiments showed that a new DNA strand is neatly knit on one side of the unwound double helix (called the leading strand), but on the other strand (called the lagging strand), the new strand forms in short pieces in a "backstitching" type of movement because of the one-way enzyme. Each piece starts with a short RNA primer, which is eventually replaced with DNA. The segments of new DNA are joined later.

Backstitching solved the problem of the directionality of the DNA molecule but raised yet another dilemma. When the backstitching enzyme (DNA polymerase) reaches the end of the parental strand against which it knits a new strand, it must somehow replace the RNA primer that started the new strand. Unless it can do this, a chromosome will shorten very slightly with each DNA replication (Figure 5.4). The situation is a little like crocheting, in which an extra stitch must be added at the end of a row to keep the rows all the same length. (Simple microorganisms such as bacteria and archaea escape the problem because their DNA is circular.) James Watson wrote about this "end replication problem" in the journal *Nature New Biology* in 1973. But, in the former Soviet Union, a young researcher had already addressed it.

THE HAYFLICK LIMIT MEETS THE END REPLICATION PROBLEM

In 1966, Alexey Olovnikov was a postdoctoral researcher at the Gamelaya Institute of the Academy of Sciences of the USSR, which was nestled in foliage near the Moscow River. Olovnikov had recently earned his doctorate in immunology, and, like Hayflick, he was investigating a viral cause of cancer.

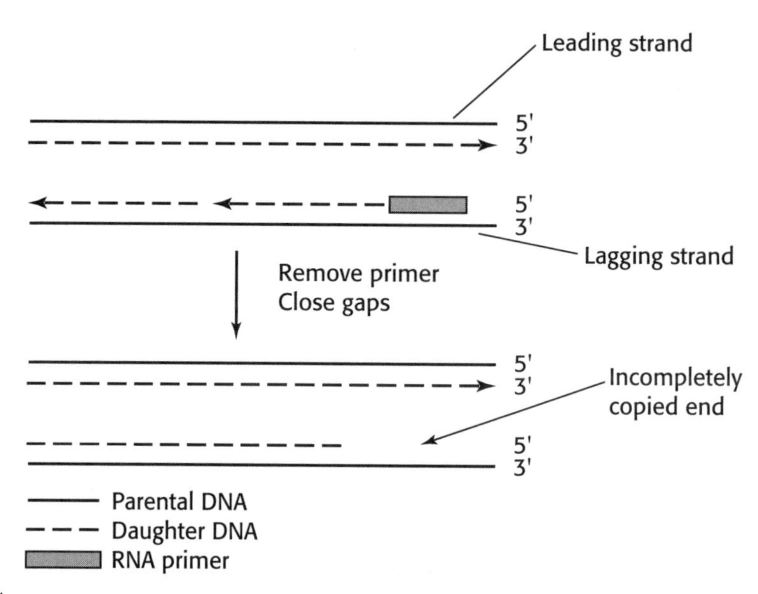

FIGURE 5.4

The end replication problem. Because of the antiparallelism of the DNA molecule and the fact that DNA polymerase can only create a new strand in one direction, some DNA at the 5'end is unreplicated, shortening a chromosome slightly with each replication.
Redrawn from Harley, C. Telomere loss: mitotic clock or genetic time bomb? Mutation Research *1991;256:271–282.*

The idea that chromosomes shrink from their ends with each round of DNA replication came to Olovnikov on a picturesque fall afternoon. He had just left Moscow University, where he'd heard a noted cell biologist describe Hayflick's work. It rang a bell. Olovnikov recalled hearing of Hayflick several months earlier in a group discussion at the Institute, but had forgotten about it. This time, "I was thunderstruck by the novelty and beauty of the Hayflick limit," Olovnikov recalls.

It took him an hour to dawdle to the metro station, absorbed in contemplating what the cellular replicometer might actually be. "I thought and thought of what this phenomenon means, how to explain that cells somehow count and remember the number of divisions already performed. I was in a Moscow metro station, and heard a deep tone of a train coming out from the tunnel. I imagined the DNA polymerase as a big construction (vehicle) moving along the DNA molecule—the tracks—like a metro train. And I thought that this polymerase cannot begin to copy from the very end, since there is a dead zone between the end of the polymerase molecule and its catalytic center, just as there is also a dead zone between the end of a carriage standing at the end of a subway's platform and the nearest door for entrance into the railroad car," Olovnikov says.

Olovnikov's "serendipitous underground brainstorm" spawned the idea that chromosomes shrink from their ends with each round of DNA replication. "Such a hypothesis looked very strange at the time, when even the fact that DNA in eukaryotic chromosomes has a linear form was a subject of discussion," Olovnikov recalls. He thought about his idea now and then over the next few years and mentioned it to colleagues. He corresponded with Hayflick, who sent him unpublished data. Eventually three senior researchers at the Institute submitted Olovnikov's "strange theory" to *Doklady* (the local equivalent of *Proceedings of the Academy of Sciences*), which published it in 1971. A year later he presented his hypothesis at the ninth International Congress of Gerontology in Kiev, traveling on vacation time because his director thought it unnecessary for him to give a talk—publishing an abstract was sufficient.

And so at the meeting in Kiev, Alexey Olovnikov presented his "theory of marginotomy." "In explaining the Hayflick effect, I proclaimed that the telomere shortening could serve as a counting mechanism, which like a molecular bookkeeper counts the number of cell doublings already performed. I called the ends of the DNA double helix the Achilles' heel, but that Nature, like a resourceful inventor, is able to turn even that weakness into a benefit" he says. He also pointed out that chromosome ends must not encode a vital protein, but instead can wear down without depriving the cell of anything essential.

At Olovnikov's talk, right up front, sat Leonard Hayflick. When Olovnikov asked Hayflick his opinion after the talk, the older man replied with a cryptic "Now you should check your theory." Both men today regret the brevity of that meeting, but cite extenuating circumstances—Hayflick was in the midst of a run-in with the KGB concerning the kidnapping of a friend, prominent Soviet dissident and gerontologist Zhores Medvedev. At the conference, Hayflick was organizing Western scientists to protect Medvedev. "I ended up confronting the KGB and couldn't talk to Alexey as long as I'd have liked," Hayflick recalls. But the Russian researcher and his "armchair speculations" impressed Hayflick.

Olovnikov again reported his theory of chromosome end shortening in 1972, in a Russian medical journal. Finally came a lengthier report in 1973 in the *Journal of Theoretical Biology,* a publication with a more international audience. Here he emphasized the key aging connection:

> Marginotomy causes the appearance, in the daughters of dividing cells, of more and more shortened end-genes, the so-called telogenes, with every new mitosis. The telogenes function as the starting points of end-replicons in chromosomes and also as "buffers," being sacrificed during successive mitoses. After the exhaustion of telogenes the cells become aged and are eliminated due to the loss of some vitally important genes localized in end-replicons. Marginotomy is therefore responsible for the loss with age of various cell clones of the body. Therefore marginotomy may be the primary cause of various disorders of age of multicellular organisms.

When *Nature New Biology* arrived in his mail, with James Watson's article describing the end replication problem, Olovnikov felt "a joy like a sportsman has, a joy that I was ahead of the great author of *The Double Helix.*" But his initial satisfaction turned to disappointment years later when researchers tackling the chemistry of telomeres at first credited Watson with pointing out the end replication problem. Although the men identified the phenomenon independently, many biologists learned of Olovnikov's 1973 paper only when Calvin Harley, now at Geron Corporation in Menlo Park, California, and Carol Greider of Johns Hopkins University pointed it out in a 1990 article in the journal *Nature.* "Olovnikov deserves the credit for linking the end replication problem to aging. But James Watson carefully defined the end replication problem, although he wrote nothing about aging," Harley says. Adds Greider, who would soon play a key role in the developing field of telomere biology, "The telomere field always cited Watson since his was the prediction of chromosome shortening people in the U.S. knew about and were testing. This is the case because we were not testing aging when we found telomerase [the enzyme that extends telomeres]."

Just a few years later, though, the story of telomeres would be blown wide open as the road of inquiry veered from the theoretical to identifying the molecular players. The explosion began with an unlikely star—a pond-dwelling, single-celled, cilia-fringed organism called *Tetrahymena thermophila.*

CLUES FROM CILIATES

The beauty of biology is that underlying principles apply to all organisms, which reflects the fact that all life on Earth ultimately descends from a common ancestor. Yet certain species are better suited to helping researchers explore particular questions than others—hence the lab rat, fruit flies, mice, Mendel's peas, and bacteria and their viruses in probing DNA. To investigate telomeres, attention turned to the ciliates.

In 1976, a postdoctoral researcher named Elizabeth Blackburn was working with *Tetrahymena thermophila* in the laboratory of Joseph Gall at Yale University. They were interested in this organism because at a key point in its life cycle, its five pairs of chromosomes shatter into 200 pieces, and then these pieces are copied over and over, yielding 10,000 mini-chromosomes—each with two ends. (The organism does this to rapidly make many copies of the genes that encode parts of ribosomes, which are vital for protein synthesis.) Blackburn and Gall realized that *Tetrahymena,* with its naturally amplified chromosomal load, offered an enriched system for probing telomeres—about 20,000 of them!

In 1978, Blackburn and Gall discovered that the telomeres of the amplified mini-chromosomes are repeats of the six-base sequence TTGGGG. They also showed, by removing chromosome ends and measuring them, that different *Tetrahymena* cells have different numbers of repeats on their chromosomes. They assessed telomere length by using an enzyme to cut chromosome ends at a point proximal to the telomere, then sizing the resulting fragments.

Over the next few years, researchers discovered similar six- or seven-base DNA sequences at the ends of chromosomes in a variety of organisms (Table 5.1). Although a given species tended to have an average number of repeats (50 to 70 for *Tetrahymena;* 2000 for humans, for example), the number varied among members of a species and

TABLE 5.1

Telomere Diversity

Species	**Telomere Sequence**
Ciliates	TTGGGG or TTTTGGGG
Flagellates and slime molds	TTAGGG
Yeast	TGTGTGG
Arabidopsis (mustard weed)	TTTAGGG
Caenorhabditis elegans (roundworm)	TTAGGC
Vertebrates	TTAGGG

even among cells of a multicellular individual. Moreover, telomeres are apparently so fundamental to complex life that one species can recognize another's telomeres. Elizabeth Blackburn, now at the University of California, Berkeley, added *Tetrahymena* telomeres to yeast cells. The yeast chromosome tips grew, adding the slightly different *Tetrahymena* telomeric sequences.

Blackburn then connected Watson's paper on the end replication problem to her investigation of *Tetrahymena*'s telomeres. If chromosomes naturally whittled down with each division, she thought, how did those in ciliates, as well as certain cells in more complex organisms, remain long? Perhaps an enzyme continually tacks the repeats onto chromosome ends. It couldn't be a polymerase, the replication enzyme that doesn't like ends. But then what was it?

At Berkeley, Carol Greider was just starting graduate school. She knew of Blackburn's work and wanted to work with her, but she had to pay her dues first. "As is typical of grad school, we did rotations, short projects in 3 labs. My second was in Liz's lab. You get a chance to look around and see how the lab works and what you might want to do. At the end of the three rotations, I went back to Liz and asked if I could join her lab, and work on this particular project that had been talked about but was not currently being actively pursued. And that was looking for an enzyme that kept telomeres long," recalls Greider. She was very happy to join the lab group. "I was even happier a year later when I found the enzyme!" she adds.

At the time, 1983, two models competed to explain the maintenance of telomere length: Blackburn's proposed enzyme, and a mechanism based on nonreciprocal recombination, in which new genetic material would be added onto a chromosome tip from another chromosome. (The recombination hypothesis would resurface later and is one focus of Greider's current research.) "It wasn't clear that you would have to have a new enzyme to maintain telomeres. That's kind of scary for a grad student! But, grad school is a good time to take a chance and play around for awhile," Greider says. That "playing around" turned out to be an intense series of experiments.

Because biology is based on chemistry, a way to dissect a living process is to carry it out apart from the organism. *Tetrahymena,* with its many telomeres, seemed a natural starting place to look for the hypothetical enzyme. So Greider made extracts of *Tetrahymena* and added a primer—a bit of DNA that telomere repeats could grab onto. She added DNA bases and labeled them with radioactive phosphorus, so that new telomeric material could be synthesized and then detected.

Soon, Greider and Blackburn indeed found something from inside the protozoan cells that tacks DNA bases, in the correct sequence, onto the primer. It was a day Greider will never forget. "The first indication that I had enzymatic activity was an experiment I did on Christmas day, 1984. But it took another nine months of experiments to reject other hypotheses. To be rigorous, you have to think about what could be fooling you, so I did several control experiments." She and Blackburn published the initial detection of enzyme activity in the December 1985 issue of *Cell,* and, following the tradition of naming an enzyme after what it does, dubbed the enzyme "telomere terminal transferase," meaning that it sticks something on the end of a chromosome.

The women spent the next two years zeroing in on the enzyme, sampling smaller and smaller fractions of *Tetrahymena* to isolate the part that promotes telomere extension. It was a little like trying to identify what part of a complex soup causes an allergic reaction. About all they knew, at first, was that the enzyme was large, but they had different notions as to what its chemical composition might be. "Liz's model was that it was all protein. I bet Liz there was an RNA component. We'd wagered a 6-pack of beer, and I'm not sure she ever paid up," recalls Greider.

Chemical tests revealed the nature of the enzyme. Greider discovered that it included a nucleic acid because adding a nuclease, which cuts up nucleic acids, abolished the telomere-extending activity. Other tests showed that the enzyme contains RNA, not DNA. Could part of the RNA component be a template, a molecular mold, for the six-base telomere sequence (Figure 5.5)? It all made sense. But like before, it wasn't enough to show what the enzyme was—it was also necessary to show what it wasn't. So they added a protease, which dismantles proteins. The enzyme activity stopped, meaning that it is both RNA and protein, a compound molecule called a ribonucleoprotein. Greider and Blackburn described the enzyme in 1987 in *Cell*, giving it a snappier name—*telomerase.* It was just about 50 years since Muller had coined the term "telomere." By 1989, they'd identified the catalytic component of telomerase as a reverse transcriptase, a chemical cousin to the same-named enzyme in retroviruses, such as human immunodeficiency virus (HIV), that enables RNA to be copied into DNA.

PROBING HUMAN TELOMERES

Identifying telomerase in *Tetrahymena* paved the way to finding it in other species, including a line of human cancer cells. This raised a compelling new question: Why would a multicellular organism control telomere length in its cells? Allowing telomeres to dwindle down to nothing would obviously doom a single-celled pond dweller. But what does telomere length have to do with the populations of cells that build tissues and organs in plants, animals, and fungi?

Robert Moyzis and his associates working on the human genome project at the Los Alamos National Laboratory decided to investigate human telomeres. Identifying the human telomere sequence, they suspected, lay in repeats. They already knew that the telomeres of *Tetrahymena* are repeats and that the human genome is packed with repetitive DNA sequences. But how would they fish out the chromosome tips? A human cell offers a mere 92 telomeres.

Creativity in science often involves combining techniques in novel ways, and that's what the Los Alamos team did. They broke open human cells to release the chromosomes and sheared them into pieces with a tiny needle. Next, they exposed the chromosome pieces to radioactively labeled telomeres from hamsters, which stuck to their human counterparts. Moyzis then discovered that these pulled-out pieces of human DNA included repeats of the sequence TTAGGG—precisely the sequence of the hamster's telomeres. But he needed further evidence that he was seeing human telomeres. Was the TTAGGG actually from a telomere, or was it just a random six-base bit from somewhere else among the human chromosomes?

To do that, he used a technique that Los Alamos was famous for, called flow cytometry. This is a way of passing chromosomes or cells through a laser beam, which illuminates and separates out structures bearing a particular fluorescent tag. Flow cytometry showed that every type of human chromosome had the TTAGGG sequence. But that still wasn't good enough: Where along a chromosome was the sequence located? Was it indeed at the ends? The researchers needed yet another tool to localize the candidate sequence. This time they turned to FISH (fluorescence in situ hybridization).

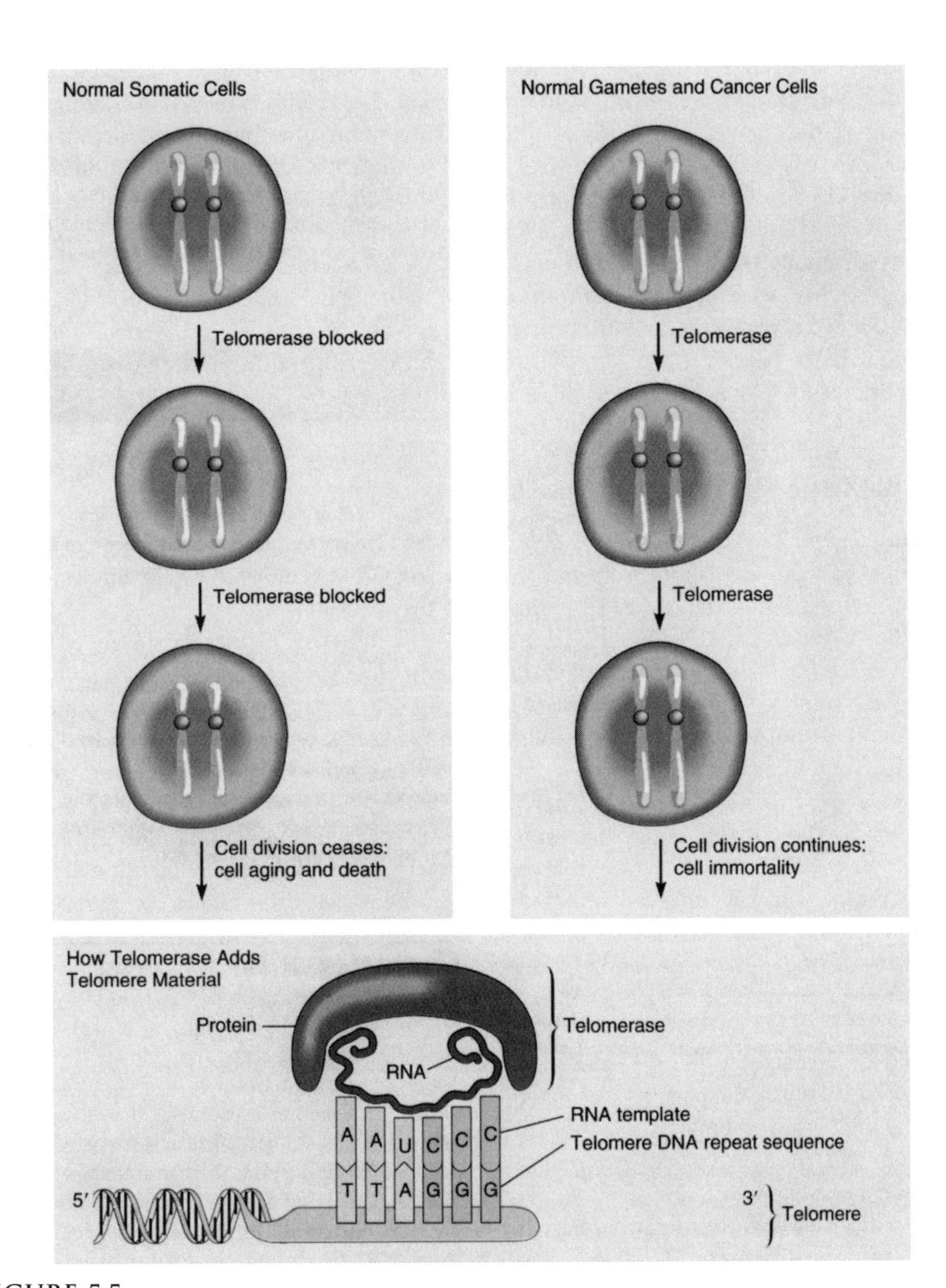

FIGURE 5.5

Telomerase adds DNA to chromosome tips by using information in an RNA template.
Lewis, R. Life, 3rd ed. Boston: McGraw-Hill, 1998:176. Reproduced with permission of The McGraw-Hill Companies.

First the researchers synthesized a DNA probe corresponding to the complementary sequence (that is, a short DNA molecule of sequence AATCCC) and chemically linked it to a molecule of the vitamin biotin. They placed the biotinylated AATCCC probe onto a microscope slide bearing human chromosomes treated to unwind the DNA. The goal was for the marked AATCCC to bind the TTAGGG telomeres. Next they added a molecule called avidin that was attached to a fluorescent dye. Avidin homes in on biotin like iron filings to a magnet, and it delivered the fluorescent tag straight to the biotin attached to the AATCCC pieces that were in turn bound to the TTAGGG sequences on the chromosomes. To boost the effect, Moyzis linked glowing antibodies to the avidin. The result was spectacular—images of human chromosomes with their tips lit up like beacons. The TTAGGG repeats—250 to 1000 of them—were indeed telomeres.

The Los Alamos researchers went on to identify the same TTAGGG telomere in more than 100 species of vertebrates, indicating that this is an ancient structure, something that must have been present in a primordial animal and spread to all its modern descendants. A dinosaur's telomeres would be the same as ours.

THE PUZZLE PIECES ASSEMBLE

With telomere structure fairly well described, attention turned to function. How, exactly, do repeated sequences stabilize chromosomes? Carol Greider, now at Cold Spring Harbor Laboratory on Long Island, was thinking about these questions.

When Greider visited her boyfriend at McMaster University in Ontario, she got to talking with the researcher in the lab next door, Calvin Harley. He knew Alexey Olovnikov's work, and told her how he had predicted, and published, the idea that chromosomes in somatic cells shorten with each cell division. Knowing now what Olovnikov didn't—the structure of telomeres and how they are probably extended—Harley and Greider teamed up to test Olovnikov's hypothesis. They measured the telomeres in cells growing in culture that had reached their Hayflick limit, in somatic cells from a fetus, and in cells from people of various ages. They saw the clear correlation that Olovnikov envisioned: The older the person who donated cells for culture, the shorter the telomeres, paralleling Hayflick's observation that the older the cell donor, the fewer remaining cell divisions.

"They did the experiments, and found the magic link," says Hayflick. The observations of three generations of telomere researchers were finally coming together, "and that convergence led to an explosion over the next 7 to 8 years. It reached to cancer biology, and was a definitive watershed in that it would show that normal cells were mortal, and cancer cells immortal. A substantial part of cell biology today rests on that distinction," says Hayflick.

It is the nature of biologists, when learning how something works, to figure out what happens when it doesn't work—or to start with the abnormal and use it to seek and explain the normal. It was clear, now, that telomere shortening was a key part of the aging process on a cellular level. What might happen to a tissue, organ, or even a person if the shortening halted unnaturally, or telomeres actually extended? As the telomere story continued to unfold in the 1990s, research would turn to questions of aging and cancer.

TRACKING TELOMERASE IN DIFFERENT CELL TYPES

The cells that build a human proceed through their allotted number of cell divisions at different rates. A cell lining the small intestine divides every few days, yet a nerve cell divides only a few times in decades, not even nearing the Hayflick limit by the time the organism of which it is a part perishes. Telomeres might be expected to be of different lengths in different types of cells—longer in the cells that form sperm and eggs because they must support a new individual, longer too in the stem cells that continually provide new cells throughout life, but shorter in highly specialized cells, such as nerve and muscle. What would be the case for cancer cells? The prediction was that they would continually produce telomerase to keep their tips extending, defying the Hayflick limit. But in 1990, Titia de Lange at Rockefeller University showed that the chromosomes in cancer cells have relatively short telomeres. They produce telomerase, but maintain their telomeres at a certain length. This could mean that turning on telomerase occurs later in the disease process, allowing some initial chromosome shortening. Therefore it appeared that telomerase activity alone was not enough to trigger or perpetuate cancer.

A series of experiments next probed several types of cells for telomere length and telomerase activity. The first evidence that cancer cells produce telomerase came in early 1994, from Calvin Harley and Sylvia Bacchetti at McMaster University. Harley had by then moved to Geron Corporation and, with a team led by Jerry Shay, Woodring Wright, and their associates at the University of Texas Southwestern Medical Center in Dallas, detected telomerase activity in a dozen types of human tumors, but *not* in 50 samples of normal cells.

It was exciting enough to find telomerase turned on in cancer cells, but the research teams discovered something even more intriguing, and potentially of great clinical utility—a correlation to stage of disease. "We've looked at more than 1000 human primary cancers, and 90% of them have telomerase activity. There is a good correlation with the stage of disease. The more advanced the cancer, the higher the telomerase activity," says Jerry Shay. For example, 68% of women in the earliest stage of breast cancer have telomerase in the affected cells, but 98% of women in more advanced stages show the enzyme activity.

For acute myelocytic leukemia, high telomerase level correlates to poor outcome. "This is a very aggressive disease. For those with high telomerase, one year later, with treatment, 20% are alive. But for the 20% of patients who have low levels of telomerase, 40% are alive two years later. Therefore, telomerase activity may be a prognostic indicator of outcome. Patients want to know, 'how long do I have, doc?,' and the doctor will be able to say, 'because your telomerase level is low, there is a 40% chance you will be alive in two years,'" Shay says.

Table 5.2 lists telomerase levels in various types of cancer. In complex tissues and organs, telomerase measurements and correlations may help to narrow down a diagnosis or render a prognosis. For example, in testicular cancers, the percentage of tumor cells that synthesize telomerase varies with cell type, such as germ cells, supportive cells, the linings of tubules, or abnormal growths of embryonic tissues. In esophageal cancer, increasing telomerase production parallels disease progression. Adenocarcinoma often follows a precancerous condition called Barrett's esophagus. One study found that 70% of early Barrett's esophagus cells produce telomerase, as do 90% of cells in a later stage. By the time adenocarcinoma has developed, all the cells manufacture telomerase. But these studies tend to track only a few dozen patients, and Carol Greider cautions that information on many more patients is necessary to determine the accuracy of measuring telomerase levels in monitoring cancer progression. Still, incorporating a telomerase assay into a cancer differential diagnosis workup is attractive, because studies indicate that the enzyme is as likely to be measureable in fine needle aspiration samples as it is in biopsies, and it can also be detected in body fluids.

Telomerase activity in the few types of somatic cells that keep it turned on is of great interest too. Titia de Lange found that certain lymphocytes normally produce telomerase.

TABLE 5.2

Telomerase Activity as a Biomarker of Cancer

Type of Cancer Activity	Percentage of Tumors Examined with Telomerase
Central nervous system lymphoma	83%
Esophageal	70% metaplasia (preprecancerous)
	90% dysplasia (precancerous)
	100% adenocarcinoma (cancerous)
Cutaneous T cell lymphoma	86%
Gallbladder	73% in gallbladder
	40% in bile duct
Gastrointestinal stromal	24% primary (first site)
	100% metastatic (spread)
Testicular	Different levels depending on cell type
Upper urinary tract	98%

And Harley, in experiments reminiscent of Hayflick's work, chronicled declining telomerase activity in the same cell types over time, finding that telomeres in liver cells from a fetus were longer than telomeres from related cells in umbilical cord blood, which in turn were longer than telomeres in stem cells in adults. "Stem cells that show renewal activity, such as blood cells, skin cells, and cells in the crypts of the small intestine, show weak telomerase activity, but the telomeres do get shorter in older individuals," explains Shay. Apparently these cells make telomerase, but not in sufficient quantities to keep the tips long.

Harley also sampled lining cells from various arteries and veins and grew them in culture, finding again that telomeres shorten as the number of cell divisions climbs. Interestingly, he found that divisions occur more often in lining cells from blood vessels that are under the greatest blood pressure. Researchers at Geron are exploring the idea that atherosclerosis arises when high blood pressure prompts blood vessel lining cells to divide too often and they build up, contributing to blockages.

As Harley and Shay's groups evaluated telomerase activity in various cell types, Carol Greider was still dissecting the composition and structure of the complex ribonucleoprotein. In 1995 she and coworkers at Geron reported the sequence of the RNA portion of the human version of the molecule. Meanwhile, other researchers were discovering that a chromosome tip is more than a naked, dangling end that attracts telomerase. It is, instead, a meeting place of sorts for various proteins. Some bind to telomerase, others bind to the telomere, and still other proteins bind to these proteins, exerting an exquisite, and still not understood, control over telomere extension. Furthermore, in mammals the dangling single-stranded end might not be the only configuration. Titia de Lange and Jack Griffiths of the University of North Carolina at Chapel Hill have seen, using an electron microscope, that the chromosome end loops around, with the single-stranded 3′ end mingling with the double-stranded telomeric region. This might be a mechanism to protect the very ends from being degraded or stuck onto other chromosome tips. It might also suggest an alternate way that telomeres can be extended from a more protected position without requiring telomerase.

MANIPULATING TELOMERASE

Research to clarify the role of telomeres and telomerase in cancer cells and normal cells has echoed the systematic versus artistic paths to discovery discussed in Chapter 1. In contrast to the rather repetitious cataloging of telomerase levels and telomere lengths in various cell types have been more creative experiments that manipulate telomerase. These more artistic efforts fall into two categories: removing telomerase from cancer cells, and adding it to normal cells.

SHUTTING TELOMERASE OFF

In 1995, Greider's team blocked production of telomerase in human cancer cells growing in culture, which triggered telomere shrinkage. Instead of dividing continuously, these cells slowed at about the twenty-third division, and eventually died. Work from Jerry Shay's group, published in late 1999, refined this experimental approach by using specific short pieces of nucleic acid analogs (oligonucleotides) to inhibit telomerase production in human immortalized breast epithelium. With telomerase activity squelched, the telomeres shortened and the cells died. Yet the response was reversible. When the researchers lifted the inhibition, allowing telomerase to be made once again, the telomeres grew. Shay sees this work as the proof of principle needed to justify development of drugs to inhibit telomerase, and several similar oligonucleotides are already in clinical trials to do just that. A variety of drugs that are different in composition, but similar in the ways in which they interact with chromosome tips, might be candidates for this potential new type of cancer therapy.

As with all cancer treatments, telomerase inhibition will have to be targeted to the affected cells or else it may elicit unwanted side effects. To examine the consequences of blocking telomerase for an entire *healthy* organism, researchers have turned to knockout technology, in which they can create a mouse that has a particular gene silenced, or "knocked out," in each cell. The gene's function is inferred by what goes awry in the animal. Chapter 6 chronicles the birth of knockout technology.

When Greider and her coworkers knocked out the telomerase gene in mice in 1998, the animals at first seemed suspiciously healthy. This tends to happen in knockout experiments and created much confusion in the early days of the technology—researchers expected life to halt when a gene was jettisoned. But life is surprisingly resilient and genomes redundant. A second gene or even a third can often stand in when the first one is knocked out. Could the apparent fitness of the first-generation knockout mice indicate that something could replace telomerase?

As Greider pondered this, and her knockout mice begat more knockout mice, she began to notice that they weren't quite so healthy after all. With succeeding generations, their robustness waned, particularly in functions that depend on highly proliferative tissues. By the sixth generation, many of the mice were infertile, their testes and bone marrow degenerated. In many cells, chromosomes were broken and fused, echoing Barbara McClintock's long-ago observations of detipped corn chromosomes and their sticky ends.

Greider also discovered that cells of the knockout mice could still become cancerous. This meant that turning off telomerase is not sufficient to halt cancer.

Greider and colleagues K. L. Rudolph and Ronald DePinho from the Dana Farber Cancer Institute also looked at elderly knockout mice—those more than 18 months old—and found that they developed only certain signs of aging, such as slowed wound healing, premature graying and hair loss, and digestive problems because cells of the frequently shed intestinal lining could no longer divide on schedule. Moreover, the aged telomerase-deficient mice had shorter life spans and an increased incidence of spontaneous cancerous tumors. The spectrum of signs and symptoms resembled Werner syndrome in humans, a condition of accelerated aging that in 1996 was linked to unusually shortened telomeres.

Taking a cue from the gradual appearance of abnormalities in highly proliferative tissues after several generations of telomere impairment in mice, Shawn Ahmed and Jonathan Hodgkin at the MRC Laboratory of Molecular Biology in Cambridge in the United Kingdom decided to look in another species. They turned to *Caenorhabditis elegans,* the tiny transparent roundworm that is to developmental biology what the fruit fly is to genetics. The researchers scrutinized 400 mutagenized lines of worms, finding 16 in which fertility clearly declined as the generations proceeded. Infertility was particularly easy to track because the worms are hermaphrodites—no need to set up crosses. Colonies that shrank over time indicated a problem with germ cells, which are maintained in their rapidly dividing state by telomerase. The mutant worms led the researchers to identify a causative protein, which, to their surprise, was *not* the reverse transcriptase part of telomerase. Instead, they found mutations in a checkpoint gene, which monitors DNA repair as a cell proceeds through the cell cycle. Sure enough, when they looked closer, the investigators found that the mutant worms could not recover from x-ray-induced DNA damage. And their cells had fused chromosomes.

Finding that a gene that controls DNA repair also has a role in telomere extension revealed something more about the molecular events at the chromosome tip. Because DNA repair occurs at a break across both strands of the DNA, in an internal portion of a chromosome, the discovery suggests that at some point the telomere resembles a double-stranded break. And so the age-old end replication problem—how the cell deals with a dangling single-stranded end—is not the only challenge to controlling telomere length. Again, a discovery catalyzed a rush of new questions.

TURNING TELOMERASE ON

At about the same time as knockout mice and mutant worms were revealing what life was like without telomerase, Harley and Shay were asking the complementary question: What happens to normal somatic cells given telomerase? To find out, they added telomerase to cultures of cell types that malfunction in certain degenerative diseases associated with aging—pigment cells from the retina that fade, causing macular degeneration; fibroblasts that cause skin to sag when their collagen-secreting ability wanes; and blood vessel lining cells that overgrow.

The effects of adding telomerase to these cells were so striking that an Associated Press report likened telomerase to Dick Clark, the perpetually youthful music show host. NBC evening news anchor Tom Brokaw announced on January 14, 1998, that "they have found a way to reverse the aging process." Not quite. But the somatic cells given telomerase indeed keep on dividing and dividing and dividing, passing through hundreds of cycles

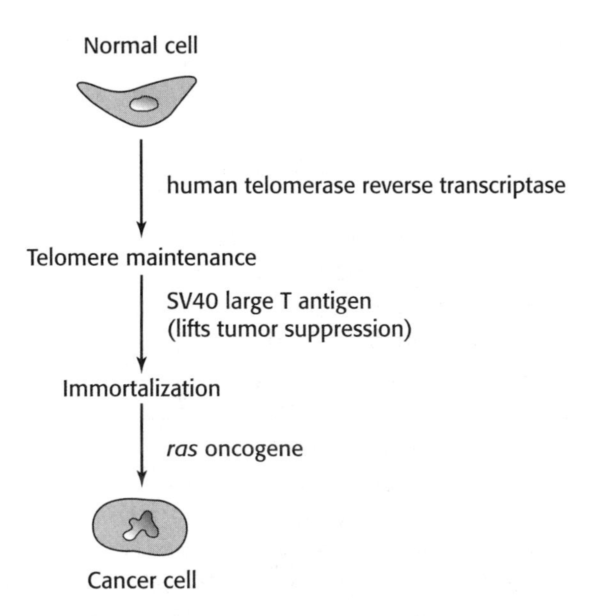

FIGURE 5.6

Making a cancer cell. Adding telomerase is not sufficient to turn a cell cancerous. It also requires an oncoprotein from the virus SV40 to turn off certain tumor suppressors, and expression of the oncogenic protein *ras*. Many different sequences of genetic events can cause cancer. Activating telomerase may be common to many, if not all, of them.
Adapted by permission from Nature *(Weitzman JB and Yaniv M. Rebuilding the road to cancer. Nature 1999;400:401), copyright 1999 Macmillan Magazines.*

of mitosis. They're still at it, and reports of their vitality surface in the news media about once a year. Even years later, the cells have not turned cancerous. Says Harley, "The cells not only have normal chromosomes, but normal controls of division. They are responding in an appropriate manner by shutting down if damaged. Cancer cells don't do that."

Adding telomerase is not sufficient to cause cancer, but it may be part of a series of changes that culminate in the cancerous state. To see how telomerase fits into this scenario, Robert Weinberg and colleagues at the Whitehead Institute for Biomedical Research and the Duke University Medical Center tried a daring experiment—they made a normal cell turn cancerous (Figure 5.6). Starting with normal human epithelium as well as fibroblasts, to show that the phenomenon is not unique to a certain cell type, they added, in sequence:

1. Human telomerase reverse transcriptase (the catalytic subunit)
2. A cancer-causing oncoprotein from the virus SV40, called T antigen, which inactivates certain tumor-suppressing genes
3. The oncogene *ras*

Sequential changes ensued. The cells began to maintain telomere lengths, exceeded the Hayflick limit to become immortal, and developed the distinctive signs of a cell turned cancerous. Headlines announced the dubious accomplishment of "making cancer cells in a test tube," but the work revealed more precisely the role of telomerase in carcinogenesis: Its presence is a crucial criterion, but one that would not cause the condition on its own.

This is valuable information for the ongoing development of telomerase inhibitors to treat cancer.

THE FUTURE: TELOMERE MEDICINE?

Clinical applications of telomere biology are eclectic and exciting (Figure 5.7). Can we block telomerase to halt cancer cells' perpetual division while preserving the enzyme's normal and necessary function in stem cells? Can we boost telomerase to grant new proliferative life to the cells whose breakdown fuels the degenerative disorders of aging? Even though telomerase modulators as drugs are years away, just understanding the dynamics of regulation of telomere length is already helping to make sense of certain clinical observations. This is the case for bone marrow transplants.

Bone marrow transplants can theoretically cure more than 60 different diseases, mostly of the blood. But in clinical reality, the procedure is very risky. Infection, rejection, and graft-versus-host disease pose constant threats. Another complication—leukemia years later—may have something to do with telomere lengths. Explains Jerry Shay, "If you look at bone marrow cells transplanted from a donor to a recipient with leukemia or another hematopoietic disease, telomeres are shorter in the cells once they are in the recipient, compared to the same cells in the donor." It seems as if transplanting cells from one body

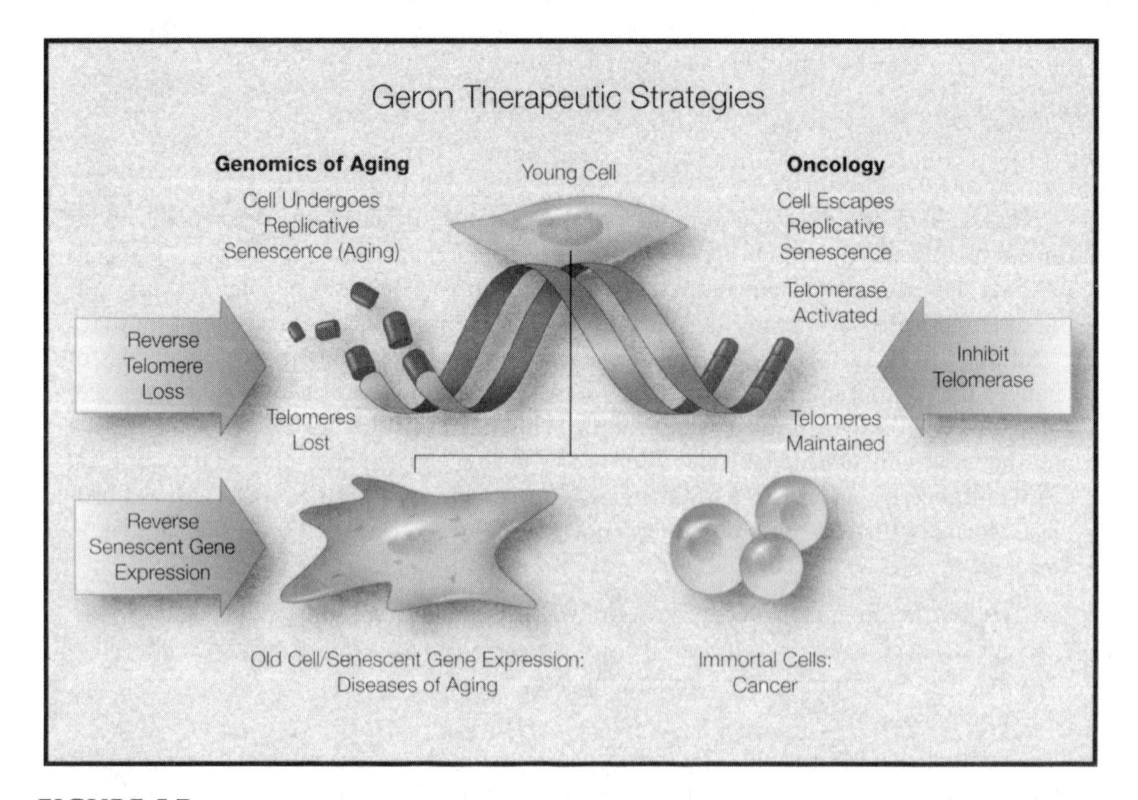

FIGURE 5.7

Telomere biology will have practical applications in understanding both cancer and aging.
Courtesy of Geron Corporation.

to another stresses them and revs up their replicometers. Plus, the fewer cells that are transplanted, the more the telomeres shrink, as if the cells share the burden of restoring the marrow and therefore age faster. Such transplanted cells accelerating through their 50 or so divisions and ceasing to divide too soon may trigger a new cancer or a recurrence. "Ten to fifteen years later, a bone marrow transplant patient may develop leukemia, because the transplanted cells did not have the proliferative capacity of a true stem cell," Shay suggests. But cells destined for transplant given a boost of telomerase first, or even genetically engineered to make more on their own, may be able to replenish marrow without becoming destabilized to the point of turning cancerous.

Exactly how new knowledge of telomere biology will be applied to the fight against cancer remains to be worked out. Researchers know that they will have to be very careful in trying to turn off telomerase because of the built-in redundancies that we do not yet understand. If a treatment blocks telomerase, something else, something even harder to fight, could become activated to keep chromosome tips long and cancer cells dividing. And that worries Carol Greider. "Clearly, if you don't have telomerase, tumor formation is curtailed. But cells have a mechanism to get around their requirement for telomerase, and that is the recombination based mechanism proposed back in 1984 to maintain telomeres, before we knew about telomerase. And so the research has come full circle." Drugs to block telomerase might give an advantage to cancer cells that can somehow recombine their way into extending their telomeres, enabling them to live forever—or until the patient dies.

But Jerry Shay is more optimistic, pointing out that such recombination is a rare event, and the cells that arise from it do not necessarily divide faster than others. He sees telomerase inhibitors as a follow-up to conventional cancer treatments, to delay the recurrences that account for most cancer deaths. "There could be a window of opportunity after surgery to have a period of anti-telomerase therapy where small pockets of residual cancer cells are still dividing," he says, adding that keeping telomeres shortening in these escapee cells could translate into added years of remission. "Telomerase inhibitors may put cancer on hold or at least slow it down, and thus provide additional and hopefully healthful years of normal life," he adds.

Applications of telomere biology seem limited only by the imagination. For example, one could take T cells from a person infected with HIV who does not yet show signs of immunodeficiency, add telomerase, and culture them. When AIDS begins and T cell supplies plunge, the patient could be reinfused with the cells, which have been nurtured to have an extended life span. Similarly, for a child with muscular dystrophy, during the early years before muscle function visibly declines, immature muscle cells could be bolstered with telomerase and grown in the laboratory. When the child's muscles begin to fail, implants of the cultured muscle cells may slow disease progression. And cells with their proliferative ability extended courtesy of telomerase would make better cell culture models than the cancerlike cells used in continuous culture today. Biotechnology companies are already selling such cells to researchers.

At the Baylor College of Medicine, investigators demonstrated the feasibility of using telomerase to revitalize implants in an elegant experiment with input from humans, cows, jellyfish, and mice. Michael Thomas, Lianqing Yang, and Peter Hornsby began with cells from a cow's adrenal gland. These cells normally produce glucocorticoid hormones that require several enzymes for their synthesis. A complex system was chosen on purpose: If the recipients could produce the hormones, then their cellular responses were intact. To the adrenal cells the researchers added genes encoding human telomerase reverse transcriptase

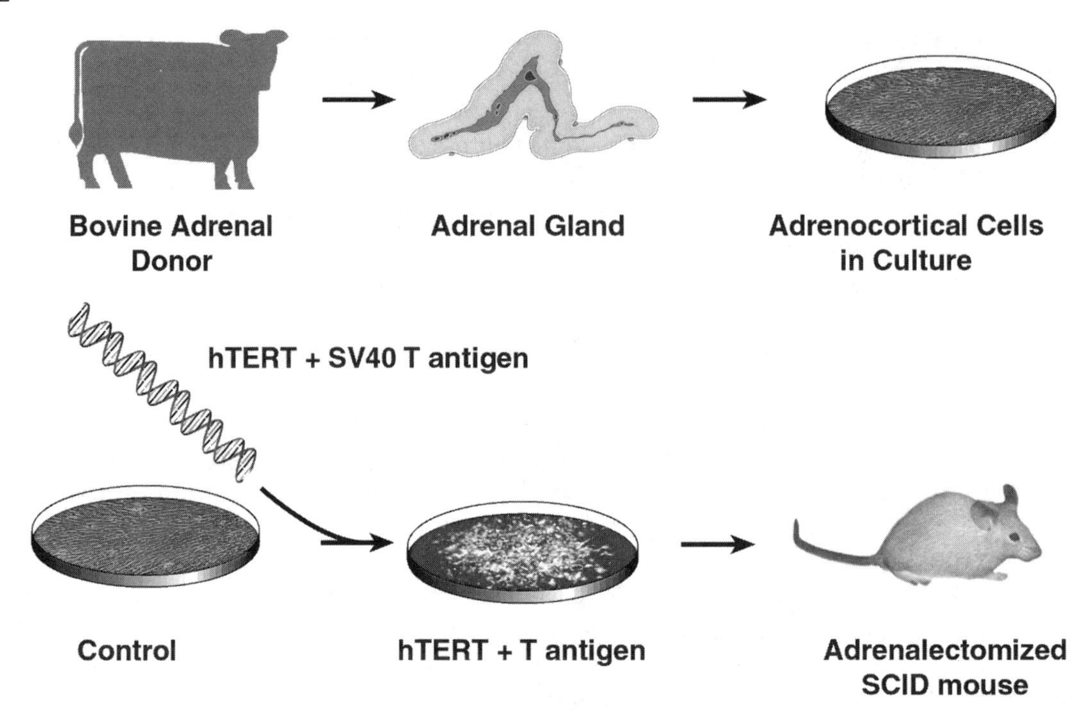

FIGURE 5.8

A proof-of-principle experiment for tissue engineering. Researchers at the Baylor College of Medicine genetically engineered adrenal cells from a cow to produce human telomerase, adding a gene from the virus SV40 that promotes immortality. The cells were encased in a polycarbonate cylinder, then implanted into mice whose adrenal glands had been removed. The mice produced the bovine adrenal hormone, and the mice's own tissues infiltrated the implant. Because the implant did not overgrow its container, it apparently did not turn cancerous, despite producing telomerase.

Reproduced with permission from Shay J. and Wright WE. The use of telomerized cells for tissue engineering. Nature Biotechnology *2000;18:22–23.*

(hTERT), the aforementioned SV40 T antigen that grants immortality, and the gene for green fluorescent protein, a jellyfish contribution that would make the eventual implants glow greenly, distinguishing them from surrounding tissue. The researchers packaged the engineered bovine cells in polycarbonate cylinders, then inserted the cylinders beneath the kidneys of mice that had had their adrenal glands removed and that lacked immune systems.

Results were spectacular (Figure 5.8). The control group mice, which received no implants but had their adrenal glands removed, died, as was expected. The luckier experimental group not only survived, but also produced the cow version of the vital glucocorticoid hormone, and the implants had become lovely, functioning mosaics of bovine tissue threaded with mouse blood vessels. Yet the implants did not outgrow their capsules, indicating that they had not become cancerous, despite the telomerase boost. This was not the case for implants of cancer cells. Therefore, telomerase and immortality by themselves could cause cell proliferation without causing cancer.

Jerry Shay and Woodring Wright called the experiment "a first step in documenting that the introduction or activation of telomerase in normal cells is likely to have many applications and a major impact for the future of medicine." Shay and Wright dubbed the implant cells "telomerized," thereby converting a noun of half a century into a verb. It is a vibrant term that will probably be heard again as tissue engineering joins, or even ultimately replaces, such other medical approaches as prostheses, surgical repair, and transplants.

The list goes on and on for possible applications of this branch of biology that began in the 1930s with observations of sticky chromosomes. Yet, almost paradoxically, Shay insists we are just beginning to fathom the power of understanding, and manipulating, the cell division replicometer. "This is a wide open field now, and we are at the starting block. We are in a position to begin to ask questions."

And that is the very essence of science—asking new questions.

REFERENCES

Ahmed, Shawn, and Jonathan Hodgkin. "MRT-2 checkpoint protein is required for germline immortality and telomere replication in *C. elegans*." *Nature* 403:159–163 (2000).

Blackburn, Elizabeth H., and Joseph G. Gall. "A tandemly repeated sequence at the termini of the extrachromosomal ribosomal RNA genes in *Tetrahymena*." *Journal of Molecular Biology* 120:33–53 (1978).

Bodnar, Andrea G., et al. "Extension of life-span by introduction of telomerase into normal human cells." *Science* 279:349–352 (1998).

Chang, Edwin, and Calvin B. Harley. "Telomere length and replicative aging in human vascular tissues." *Proceedings of the National Academy of Sciences* 92:11190–11194 (1995).

Chen, Jiunn-Liang, et al. "Secondary structure of vertebrate telomerase RNA." *Cell* 100:503–514 (2000).

Feng, Junli, et al. "The RNA component of human telomerase." *Science* 269:1236–1241 (1995).

Gasser, Susan M. "A sense of the end." *Science* 288:1377–1379 (2000).

Greider, Carol W., and Elizabeth H. Blackburn. "Identification of a specific telomere terminal transferase activity in *Tetrahymena* extracts." *Cell* 43:405–413 (1985).

Greider, Carol, and Elizabeth Blackburn. "The telomere terminal transferase of *Tetrahymena* is a ribonucleoprotein enzyme with two kinds of primer specificity." *Cell* 51:887–898 (1987).

Greider, Carol, and Elizabeth Blackburn. "A telomeric sequence in the RNA of *Tetrahymena* telomerase required for telomere repeat synthesis." *Nature* 337:331–337 (1989).

Griffith, Jack D., et al. "Mammalian telomeres end in a large duplex loop." *Cell* 97:503–514 (1999).

Hahn, William C., et al. "Creation of human tumor cells with defined genetic elements." *Nature* 400:464–468 (1999).

Harley, Calvin B. "Telomere loss: Mitotic clock or genetic time bomb?" *Mutation Research* 256:271–282 (1991).

Hayflick, L., and P. S. Moorhead. "The serial cultivation of human diploid cell strains." *Experimental Cell Research* 25:585–621 (1961).

Herbert, B.-S., et al. "Inhibition of human telomerase in immortal human cells leads to progressive telomere shortening and cell death." *Proceedings of the National Academy of Sciences* 96:14276–14281 (1999).

Herskowitz, I. J., and H. J. Muller. "Evidence against the healing of x-ray breakages in chromosomes of female *Drosophila melanogaster*." *Genetics* 38:653–704 (1953).

Kim, Nam W., et al. "Specific association of human telomerase activity with immortal cells and cancer." *Science* 266:2011–2015 (1994).

Lee, Han-Woong, et al. "Essential role of mouse telomerase in highly proliferative organs." *Nature* 392:569–574 (1998).

McClintock, Barbara. "The behavior in successive nuclear divisions of a chromosome broken at meiosis." *Proceedings of the National Academy of Sciences* 25:405–416 (1939).

Moyzis, Robert K. "The human telomere." *Scientific American*, August 1991, 48–55.

Olovnikov, A. M. "A theory of marginotomy." *Journal of Theoretical Biology* 41:181–190 (1973).

Olovnikov, A. M. "Telomeres, telomerase, and aging: Origin of the theory." *Experimental Gerontology* 31:443–448 (1996).

Rudolph, K. L., et al. "Longevity, stress response, and cancer in aging telomerase deficient mice." *Cell* 96:701–712 (1999).

Shay, Jerry. "At the end of the millennium, a view of the end." *Nature Genetics* 23:382–412 (1999).

Shay, Jerry, and Woodring E. Wright. "The use of telomerized cells for tissue engineering." *Nature Biotechnology* 18:22–23 (2000).

Thomas, Michael, et al. "Formation of functional tissue from transplanted adrenocortical cells expressing telomerase reverse transcriptase." *Nature Biotechnology* 18:39–42 (2000).

Watson, James. "Origin of concatemeric T7 DNA. *Nature New Biology* 239:197–201 (1972).

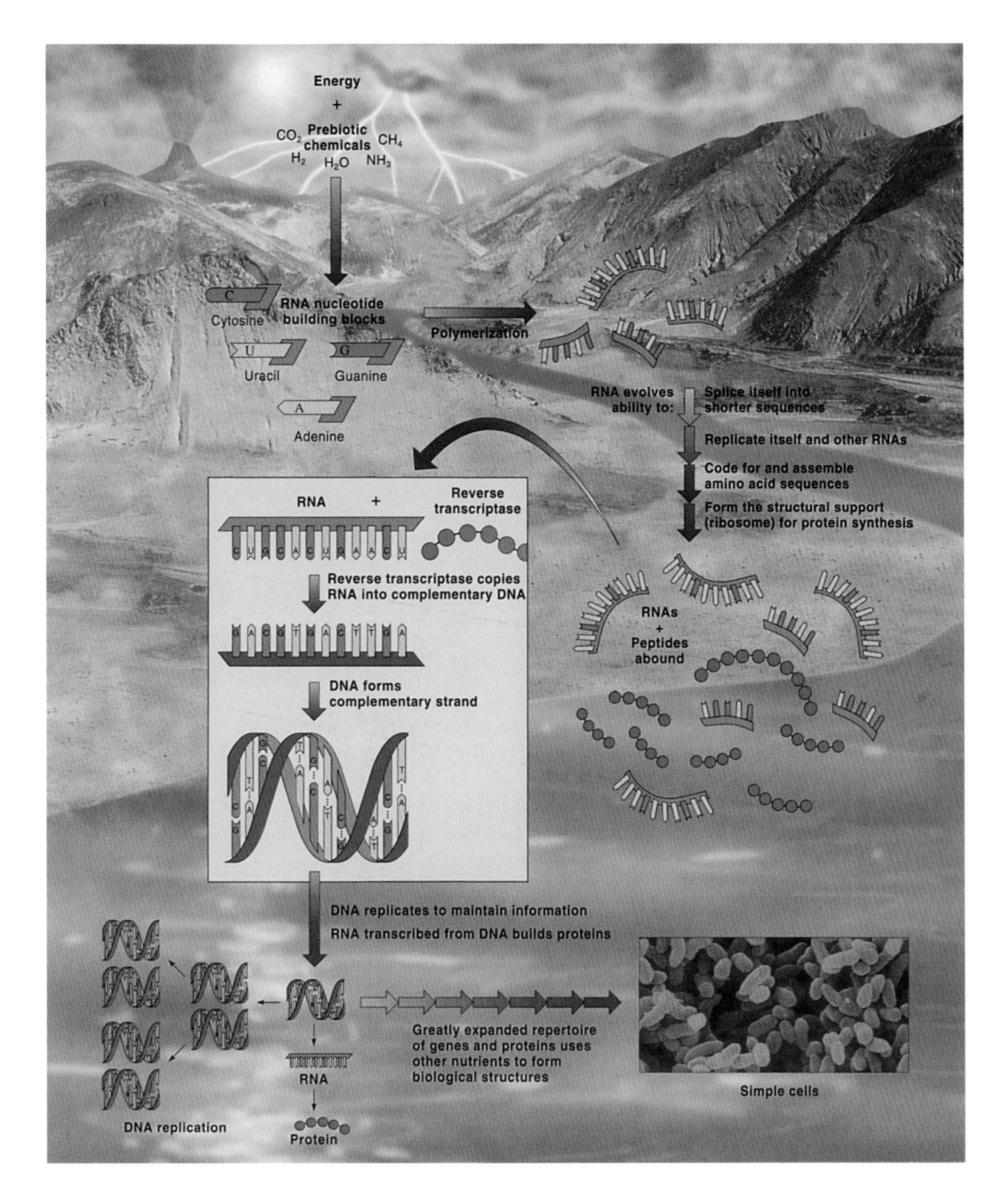

COLOR PLATE 1

The RNA world was probably not a starting point, but instead a culmination of a pre-RNA world where many types of chemicals that could have been the precursors of biochemicals formed and perhaps interacted. Eventually, DNA and protein took over some of the functions that may have originated in RNA or a similar type of molecule.

Lewis R. Life. 3rd ed. Boston: McGraw-Hill, 1998:356. Reproduced with permission of The McGraw-Hill Companies.

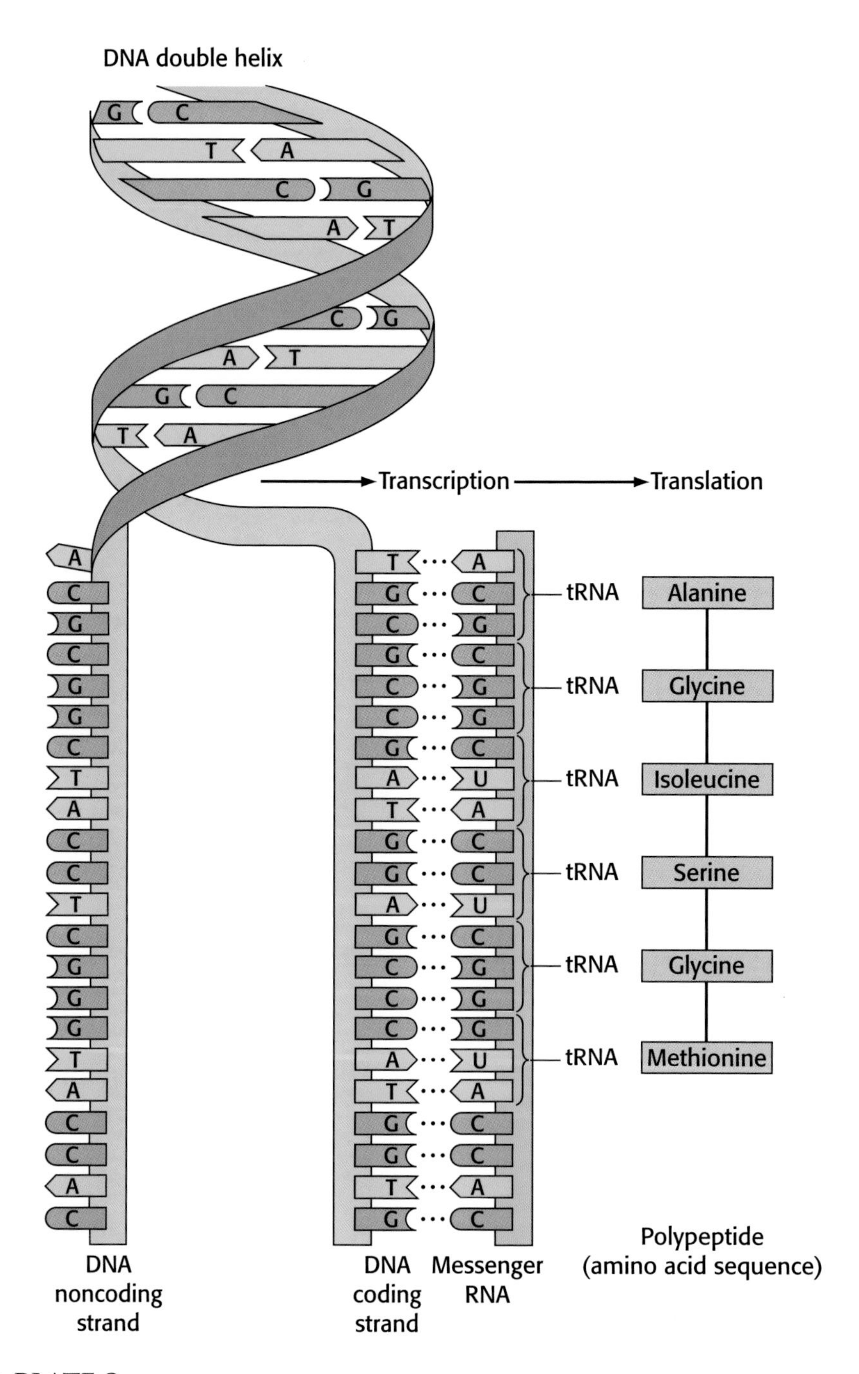

COLOR PLATE 2

The flow of genetic information. One side of the DNA double helix is transcribed into messenger RNA (mRNA), to which the appropriate transfer RNAs (tRNAs) bring amino acids. A protein is built as the aligned amino acids join.

Adapted from Lewis R. Human genetics: concepts and applications. *3rd ed. Boston: McGraw-Hill, 1999. Reproduced with permission of The McGraw-Hill Companies.*

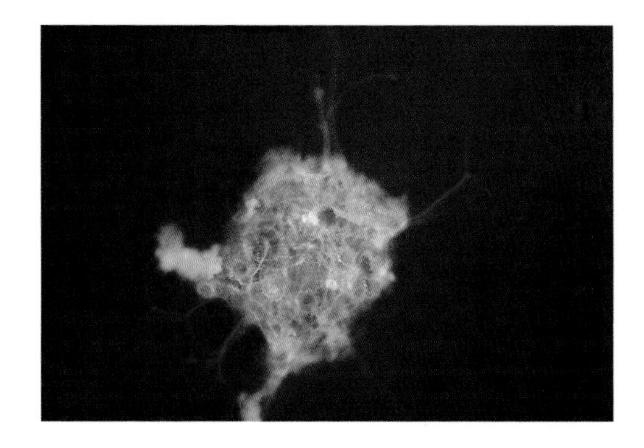

COLOR PLATE 3

Neural stem cells form balls called neurospheres when cultured.
Courtesy of Dennis A. Steindler, The University of Tennessee, Memphis.

2 Routes to Obtaining Clones of Stem Cells

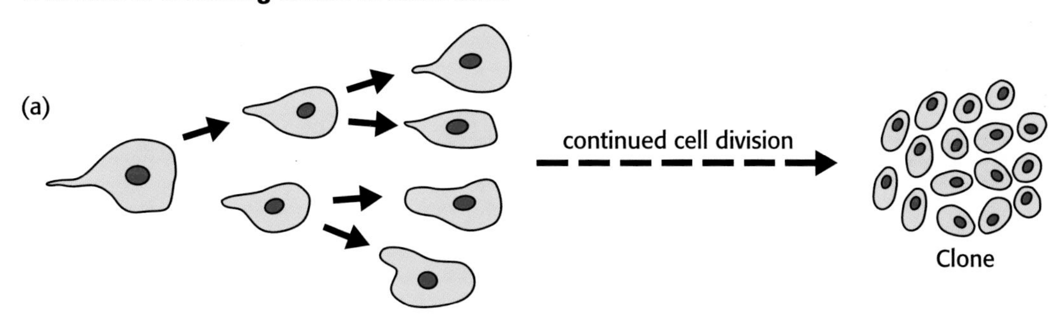

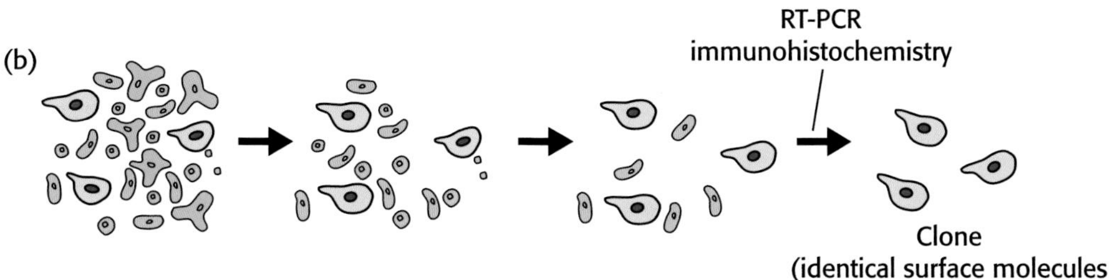

COLOR PLATE 4

Researchers observe clonally related groups of cells in two ways: (a) by starting with one cell and visually tracking its descendants and (b) by examining the surfaces of cells to demonstrate that they are identical.

COLOR PLATE 5

Monarch butterfly larvae typically eat the leaves of milkweed plants, but avoid those coated with corn pollen.

COLOR PLATE 6

"Yellow rice," unlike its unaltered counterpart, produces β-carotene. It also manufactures twice the normal amount of iron.
Courtesy of the Botanical Congress, St. Louis.

CHAPTER 6

STEM CELLS: THE SCIENCE OF SELF-RENEWAL

Within the nucleus of nearly each cell of a many-celled organism lies, encoded in about 2 meters of DNA, all the information needed to construct that particular organism. If the cell is from a human, those biochemical blueprints specify thousands of different proteins, as well as the RNA molecules that can translate the instructions. But, like a huge library from which only a few books are borrowed, most cells access only a small fraction of the volumes of information sequestered within them.

Although all cells use genes that provide the basic functions of life—so-called housekeeping genes—at the same time they pick and choose among the more specialized instructions. A bone cell continually produces collagen protein, burying itself in concentric rings of mineral-studded matrix as it utilizes that particular gene's instructions over and over, like a library visitor who repeatedly borrows The World According to Garp. *In contrast, a muscle cell fills with fibrils of the contractile proteins actin and myosin, yet doesn't tap its ability to manufacture collagen. And a nerve cell doesn't require much collagen or myosin, instead manufacturing the parts of its inner skeleton that enable it to sprout extensions that can reach toward other cells and to produce, package, and release neurotransmitters.*

Just as a library can accommodate fans of mysteries, biographies, or science fiction, so too can the genome tucked into each cell accommodate the specialized fate of that cell. Theoretically, any cell with a nucleus retains the potential to become anything. But certain cells, called stem cells, *are especially plastic, having not yet "decided" which developmental pathway to follow. When such a stem cell divides, it may yield two of itself, or one stem cell and another cell that goes on to specialize into a differentiated cell type. Not all stem cells are created equal, in terms of their potential. A stem cell that can specialize into virtually any type of cell is termed* totipotent, *a cell with a more restricted range of fates is* pluripotent, *and a stem cell already on the road to forming a particular cell type is referred to as* multipotent *or as a* progenitor cell. *If a totipotent cell is likened to a large public library, with many choices of books, then a pluripotent cell would be a science library, and a progenitor cell perhaps the astronomy section of that library (Figure 6.1).*

What would happen if we could tap the potential in a stem cell and direct it?

FROM BONE MARROW TO BRAIN MARROW

If we could locate and harness the body's stem cells, then perhaps we could learn how to guide them toward specific fates by providing appropriate biochemical signals, thereby creating "spare parts." Someone suffering from a spinal cord injury might donate a few

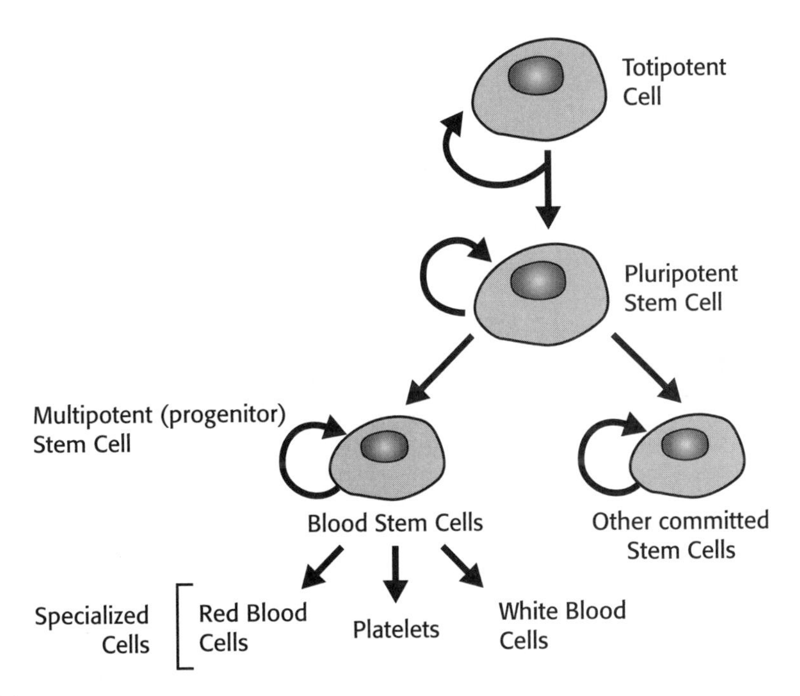

FIGURE 6.1

A totipotent stem cell retains the developmental plasticity to become any cell type. When it divides, it can yield another totipotent cell or a less specialized, pluripotent cell that can fulfill a subset of all possible developmental fates. A progenitor cell is even further restricted, such as a stem cell in the blood that can give rise to any of the three major types of the blood's formed elements.
National Institutes of Health.

cells that, in laboratory culture, could be nurtured to grow into needed neural tissue that would be grafted at the site of the damage. Universally compatible neural stem cells might be introduced into the brains of people with neurodegenerative conditions such as Alzheimer's or Parkinson's disease, where they might make connections that restore some function. Stem cells might help heal a heart attack or seed a malfunctioning pancreas with cells able to secrete insulin, helping a person with diabetes from within. Already a person with a blood cancer can have her abnormal cells replaced with her own healthy cells that have been precisely picked from her bone marrow, expanded outside the body, and then infused. Stem cells coaxed to specialize in the laboratory can also be used for drug testing and in basic research (Figure 6.2). Using the body to heal itself already has a name—regenerative medicine—and an underlying technology, tissue engineering. And with an aging population, this approach may just well be the wave of the future.

Stem cell technology is a type of tissue engineering that captures the not-yet-expressed capabilities or reawakens the already-silenced capabilities of cells. Such cells come from two sources. Embryonic stem (ES) cells are cultured from cells of embryos that retained their potential simply because they hadn't developed far enough to shut off many of their genes. The second source of stem cells consists of the packets of cells in our bodies that serve as natural reservoirs for growing, repairing, and healing. Many types remain to be discovered. The best studied "adult" stem cells are in umbilical cord blood and bone marrow.

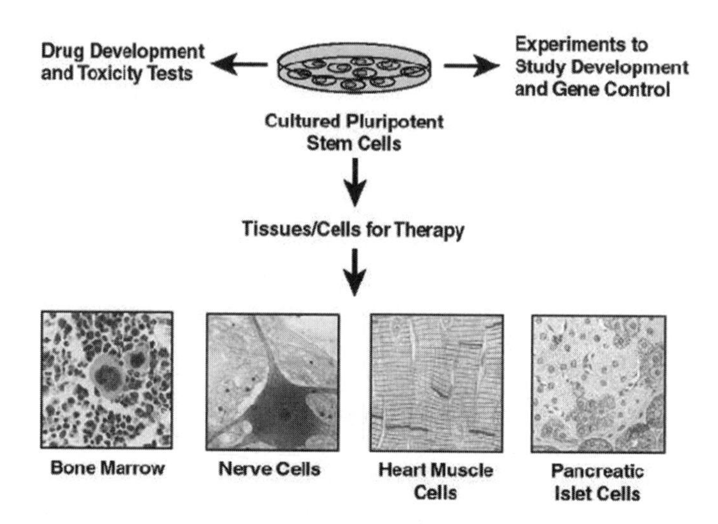

FIGURE 6.2

Stem cells stimulated to divide and differentiate in the laboratory can be used for *in vitro* drug testing, to study tissue formation and development, and for "spare parts," a new field called regenerative medicine.
National Institutes of Health.

These multipotent cells can repopulate the complex mixture that is blood and are turning out to be more plastic than anyone imagined. Under certain conditions, certain bone marrow stem cells can travel and divide, specializing into the fibroblasts of connective tissue, cartilage, fat, bone, liver, neurons, or muscle. That is, the bone marrow's multipotent cells may be more "multi" than we had thought, and so, too, the body's ability to selectively renew itself.

Even the brain, whose neurons have mostly ceased dividing, harbors a storehouse of neural stem cells that eluded detection for many years simply because it is a mere smattering of cells hugging a fluid-filled ventricle. A neural stem cell can give rise to neurons as well as glia, the other cells of the nervous system. And like some stem cells in bone marrow, certain neural stem cells can, under certain conditions, detour into entirely different developmental pathways. Destroy a mouse's bone marrow, for example, and neural stem cells start over and supply the needed blood cells. So too can stem cells in muscle divide to yield blood cells. It is as if the reader who repeatedly takes out *The World According to Garp* suddenly finds that his favorite book has become *Moby-Dick.*

"These cells have more potential than we ever granted them. The brain is not so special a place after all," says Dennis Steindler, a professor of anatomy and neurobiology at the University of Tennessee. His research team cultured human adult neural stem cells, which have come to be called "brain marrow." In a lab dish, the cells form balls called neurospheres (Color Plate 3). Steindler likens brain marrow to a pile of bricks left behind at the site of a finished building, ready to repair damage should it become necessary. But considering the prevalence of neurodegenerative disorders, the brick stack often isn't sufficient. At least two biotechnology companies have developed methods to isolate adult neural stem cells and maintain them in the laboratory, with an eye on a different sort of potential than the developmental one—a possible $150 billion-per-year drug market.

The beauty of regenerative medicine courtesy of adult stem cells is that it borrows from normal physiology. "Every adult human has these stem cells, which are used in normal tissue turnover. If you break your arm, the body recruits mesenchymal cells to form osteoblasts and repair the fracture. These cells proliferate and engraft at the site and differentiate following a normal pattern," explains Daniel Marshak, chief scientific officer at Osiris Therapeutics, a Baltimore-based biotechnology company developing stem cell replacements for bone marrow, cartilage, heart muscle, bone, and tendons. Elsewhere, stem cell implants have strengthened the brittle bones of a few youngsters with osteogenesis imperfecta, an inherited disorder of collagen.

According to the media, stem cells burst onto the scene at the end of 1998. *Science* magazine bequeathed their "breakthrough of the year" designation to stem cells in 1999, officially dating the debut of the field to 1981, when mouse ES cells were isolated. But developmental biologists—once called embryologists—have been investigating stem cells for more than a century. German experimental pathologist Julius Cohnheim proposed in 1867 that bone marrow sends out emissary cells to distant sites, where they become the fibroblasts required to secrete the collagen that closes wounds. Cohnheim injected an insoluble dye into the bloodstreams of laboratory animals, then observed dyed fibroblasts at wound sites, inferring that the blood, and therefore ultimately the bone marrow, delivered the cells that became fibroblasts. On the embryonic front, although human ES cells were indeed first isolated in 1997 and the work published in 1998, the roots of using cells from embryos to replenish specialized tissues actually go back to observations made a half century ago.

A CELL'S POTENTIAL GONE HAYWIRE: TERATOMAS

It is the stuff of talk shows. An 800-pound mass in a woman's abdomen contains teeth and hair and a jumble of tissue types. A TV program called "The World's Most Frightening Videos" features a man with a second face—an extra nose and mouth that move in unison with his normal ones. The medical literature offers equally strange reports: a newborn girl with "the underdeveloped lower half of a human body" in her lower back, a young man with a "large greenish mass replacing the left eye" that contains pieces of an embryo's cartilage, fat, muscle, intestine, and brain. In another case, an x-ray clearly shows a perfect set of molars embedded in what appears to be a jaw—in a woman's pelvis. And about a dozen reports describe young men who reached puberty early, then grew the distinctive cell layers of an embryo in their pineal glands, located in their brains! More common are pregnant women who carry not an embryo or fetus, but disorganized masses of specialized tissues, delivering amorphous lumps that sometimes include teeth and hair. Similar growths arise in men, in certain testicular tumors.

A tumor that includes specialized tissue cropping up in strange places is called a *teratoma.* It usually starts out as a germ cell—a cell destined to become sperm or egg—that has gone astray, its developmental program activated at the wrong time and place. Or, a teratoma might begin as a cell from a very early embryo that somehow lost its way and persisted as a body developed around it, much later reactivating its developmental program to form embryo bits in an adult. Many a teratoma also includes cancer cells and is then called a *teratocarcinoma.*

Whatever the initial cell source and trigger, the confused founding cell of a teratoma goes about forming the three tissue layers of the embryo: the innermost endoderm; the middle layer, or mesoderm; and the outer ectoderm. Many biologists working with many types of animals have meticulously tracked the development of these three fundamental tissue layers in normal embryos, creating "fate maps" that predict what each layer will become. Ectoderm begets linings (epithelia) and nerve tissue, mesoderm produces muscle and connective tissues (bone, cartilage, and blood), and endoderm gives rise to the digestive and respiratory tracts. These designations were thought to be immutable. The recent discovery that neural stem cells (ectoderm) can become blood (mesoderm) defies a century of established embryological fact.

A teratoma, however, obeys the three-layer rule, with its layers yielding the structures that they would normally become, but not in any form as well organized as an embryo. The hodgepodge nature of a teratoma is especially striking in ovaries and testes. Here, structures that accompany an embryo, such as a yolk sac and amniotic cavity, arise right next to such mature tissues as parts of the epidermis, liver, and intestine. One teratoma, like a deranged biological mosaic, consisted of epithelium surrounding a nest of red blood cells next to blobs of cartilage and connective tissue, all surrounded by hardened skinlike cells.

Teratomas are quite rare in humans, but they are common in certain strains of mice. And with the teratomas of mice lies the genesis of human ES cell technology.

STRAIN 129 LEADS TO IDENTIFYING STEM CELLS

Leroy Stevens arrived at the Jackson Laboratory in Bar Harbor, Maine, in 1953, fresh from earning his doctorate in developmental biology (Figure 6.3). The laboratory opened in 1929 with the goal of studying mice to better understand and develop treatments for cancer. Today the lab is widely known as a storehouse for mice of a dizzying variety of genetic backgrounds, a valuable resource back then and an absolutely essential part of biomedical research today.

Like many scientific discoveries that come to found fields, Stevens' work on stem cells happened serendipitously but required an eye trained to spot the unusual, in the tradition of Louis Pasteur's famous observation that "chance favors only the prepared mind" (see Chapter 1). When molecular biology was still very much in its infancy, embryologists spent their days scanning dozens, hundreds, even thousands of animals, becoming instantly alert when a fly, frog, or mouse stood out from the crowd. So it was in the Jackson Lab. Stevens, who today lives in Maine near the lab, following a serious stroke in 1990, recalls his first, and "crazy," assignment. "The founder of the lab had gotten a grant from a tobacco company, and they wanted him to show that it wasn't tobacco that was the problem—it was the paper in cigarettes!" So Stevens dutifully exposed many mice to all sorts of cigarette components in search of the one that would make them sick. Thinking the work quite insane, not to mention tedious, he was understandably thrilled when, one fall day, he spotted a teratoma in one of his experimental subjects. "I was pretty good at reading slides, and I spotted a teratoma in a mouse. I knew what it was immediately." So he wrote a grant proposal to investigate the bizarre growths further, and the resultant funding, happily, lasted a professional lifetime. "This stuff was extremely interesting, and it sure beat studying cigarette papers!" he recalls.

The original abnormal mouse was of a particular strain, number 129, and it attracted attention because it had an enormous scrotum. "We killed it, and looked at the testes, and they had strange things inside," recalls Don Varnum, a long-time technician in the lab who is known for his "miraculous hands" as a rodent surgeon. Stevens described the growth as containing "chaotic mixtures of many tissue types of various degrees of maturity."

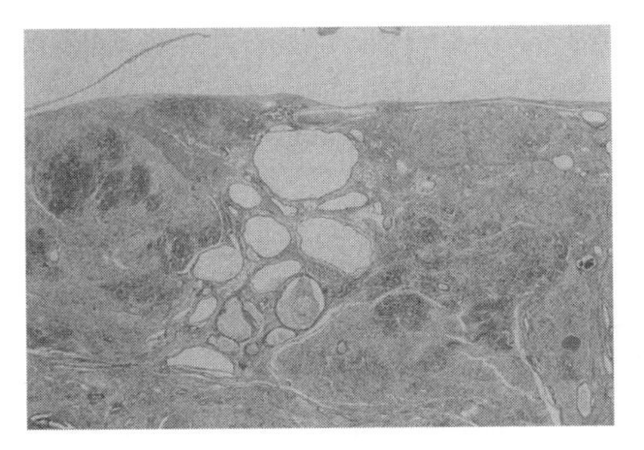

FIGURE 6.3

Leroy Stevens published exquisite photographs of the jumbles of tissue that were mouse teratomas. His work on teratomas led to identification of the two types of cells that would eventually be used to derive human embryonic stem (ES) and embryonic germ (EG) cells.
Photographs courtesy of Jackson Laboratory.

Stevens and Varnum then began looking in earnest at strain 129 mice and soon found others with the strange growths. Varnum recalls that they "drifted into teratoma research," whereas Stevens remembers it more as being "catapulted, with great excitement." Either way, their initial inquisitiveness and desire to get away from tobacco research led them down a path that neither could have predicted.

About 1% of the strain 129 mice turned out to harbor teratomas. The growths usually consisted of differentiated cells as well as "embryonal" cells, the latter of which were unspecialized, like the cells of an embryo, and cancerous. These cells could divide continuously, unlike normal cells that typically stop dividing after a set number of times (see Chapter 5). When and if the teratoma ran out of embryonal cells, it would stop growing and dividing and just sit there, sporting its peculiar mix of hair and teeth, cartilage, and tiny tubules. This wasn't nearly as interesting to Stevens as the cancerous stage, so he thought of a way to keep a teratoma youthful—serial transplants. When a teratoma neared the stage of boring specialization, Stevens and Varnum would excise it and transfer it to another host, where they could continue to watch what it would become. "We just wanted to keep the tumor alive longer so we could study it. But it took a long time to succeed. We wanted to have a continual supply of tumors to study," says Stevens.

By passaging the teratomas through many hosts and removing them from time to time and cataloging the tissues, the researchers were able to see all that these cells could become.

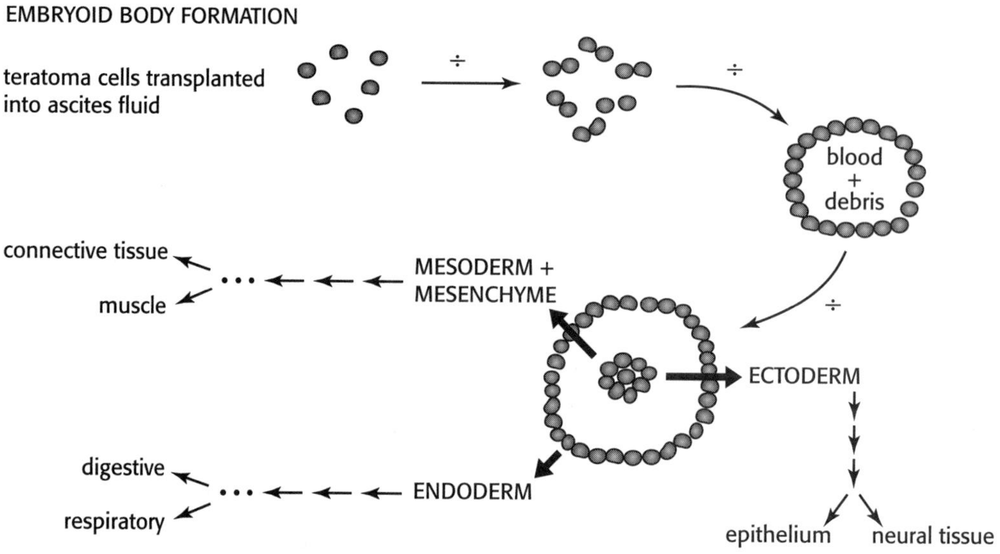

FIGURE 6.4

Teratoma cells transferred to a mouse's belly respond to factors in the ascites fluid, differentiating and interacting to form layered structures called embryoid bodies. These bodies are a little like inside-out early embryos.

Gradually, they noticed that the growths weren't as haphazard as they seemed—they were exquisitely attuned to their surroundings.

Teratomas are, histologically speaking, a mess. It is as if someone took a bunch of cells of different types, threw them in the air, and mushed them together to form a jumbled mass of tissue. But Stevens and Varnum found that when they transplanted tiny bits of a teratoma into the peritoneal cavity of a mouse, into a bath of ascites fluid in its belly, the disorganization was not as great. Something happened. A thing, more than a bunch of cells, but certainly less than an embryo, formed. It is called an embryoid body.

An embryoid body begins to form when the few teratoma cells transplanted into the ascites fluid divide to the point that they touch. The contact activates a program. The cells form a ring that at first encloses blood and cellular debris. Then a few unspecialized cells appear in the middle (Figure 6.4). The cells have taken the first step toward differentiation in that they occupy different positions with respect to each other—there is now an inside and an outside. This is precisely what happens as an embryo develops.

The cells in the middle layer of the transplanted teratoma eventually form mesoderm, as they would in a proper embryo, and mesenchyme, the tissue that can give rise to others in an adult. Cardiac muscle cells may even come together and beat in unison, forming a hint of a heart. The outer ring of cells is endoderm, the precursors of respiratory and digestive tracts. And the inner sanctum, the unspecialized cells at the center of the embryoid body, eventually become the equivalent of ectoderm. If it persisted beyond a few days, this ectoderm would give rise to nervous tissue and linings. But this somewhat organized mass trying to become an embryo has its pattern reversed, wearing its endoderm on the outside. Still, the very existence of these embryoid bodies reveals the developmental potential of a teratoma's cells, given whatever factors in the ascites fluid provide appropriate stimuli.

Back at the Jackson Lab, after a few years of what Varnum calls "playing around" with serial transplants and embryoid bodies, they added a gene called "steel" to strain 129 mice. This upped the percentage of teratoma-ridden males to 10%, making the strange growths easier to harvest. The researchers then spent several years cataloging and describing the types of tissues that make up teratomas and embryoid bodies. The papers that they churned out featured spectacular atlases, page after page of photographs that captured the temporal unfolding of the deranged development. Although the structures are not quite right, somewhat out of place and out of time, they adhere in a general sense to an inborn schedule of events as the cells interact to attempt to assemble tissues and organs. "Roy looked at thousands and thousands of mice. For tissues to develop, several tissues must be in contact, so that the cells know whether to become liver or kidney," Varnum recalls.

Although Stevens and Varnum's work was purely basic research, the interactions that they meticulously chronicled and the biochemicals from without and within that fueled them are what stem cell researchers and tissue engineers are deciphering today.

TRACING THE ORIGIN OF A TERATOMA

With a steady supply of teratomas, and a means to propagate them via serial transplants, Stevens could ask the questions that he found the classic histology books sidestepped: Where did teratomas come from? What was the cell of origin?

Pinpointing the genesis of a teratoma required a trip back in development, sacrificing mice at every step from fertilized egg onward to see just when a cell or two diverged from the masses in some way. This happened on the fifteenth day after conception—Stevens noticed that one or two cells in the tubules of the developing testes looked different. But he knew that developmental changes likely begin before cells actually look unusual to an inquisitive human. So Stevens backed up, picking cells from a part called the genital ridge in a 12-day prenatal mouse. This structure houses primordial germ cells, which develop into sperm within the coiled seminiferous tubules of the testes. In an elegant series of experiments, he and Varnum of the magical hands transplanted genital ridge cells to livers, spleens, and testes of mice. And they saw teratomas form in the testes. That meant that primordial germ cells can become the mishmash of tissues that make a teratoma.

But the results of experiments aren't always as they at first seem. In this case, it wasn't being part of the testes that coaxed the transplanted cells to form teratomas, but the temperature! The testes hang from the body in the scrotal sac, where the lower temperature is necessary for sperm to develop. "We stumbled onto this by accident. If we transplanted genital ridge into testes that were cool, or into a footpad that is cool, or into ear cartilage, which is also cool, we'd get 95 to 100 percent teratomas! But if we transplanted genital ridge into testes kept surgically in the body, so they couldn't descend, we'd get only 4 to 5 percent teratomas. If we did anything to keep the testes warm, we wouldn't get as many teratomas" Varnum says. With an ever-increasing supply of the strange growths, thanks to the new knowledge of their temperature preferences, the research continued.

In 1970, Stevens conducted a series of experiments that would profoundly affect stem cell technology a decade later, although the work is rarely, if ever, mentioned in the media version of the rise of stem cell research. Noticing that the primordial germ cells that gave rise to teratomas looked a lot like the cells of considerably earlier embryos, he transplanted cells from various stages of early strain 129 embryos into testes of adult mice, even cells from two-cell embryos. And some gave rise to teratocarcinomas! His induced growths looked identical to spontaneous teratomas, and they behaved the same; that is, cells from both natural teratomas and strain 129 early embryo cells transplanted into testes grew into embryoid bodies when transplanted into mouse bellies. Stevens called these cells from early strain 129 embryos that could support differentiation "pluripotent embryonic stem cells." But because they could give rise to cancerous as well as normal cells, they became known as *embryonal carcinoma (EC) cells.* The seeds were sown for stem cell biology.

A SURPRISE ANNOUNCEMENT CATALYZES CELL CULTURE

Over the years, many developmental biologists visited Leroy Stevens at the Jackson Lab, staying for weeks or even months to learn the techniques of mouse embryology. One such visitor was a young woman named Beatrice Mintz.

In 1973, Mintz and her coworker at the Institute for Cancer Research in Philadelphia, Karl Illmensee, borrowed some teratoma-derived cells that Stevens had been maintaining in serial transplants for eight years. These cells, if allowed to grow into a mouse, would result in an animal with a coat color called black agouti. Mintz and Illmensee mixed Stevens' black agouti cells with cells from an early embryo of a mouse destined to be brown. When the offspring were chimeric—mosaics with telltale black agouti patches—the researchers knew that the borrowed black cells, despite their teratomatous and therefore cancerous origin, could in fact support normal development.

Mintz and Illmensee saved the results of this key experiment for a surprise announcement at a meeting in 1975. Stevens, who was in the audience, was floored. "My friends and I asked her if she was going to say anything interesting or unusual—and she said no, so a lot of my friends went out sightseeing, but I stayed to hear her talk. It was then that she announced for the very first time about the chimeric mouse. It was very exciting. The work showed that the normal embryonic cells pass some kind of message to the tumor cells to stop being tumor and do something useful . . . and they did. But she was extremely secretive, which is not how it should be in scientific circles. I practically exploded," he recalls.

The creation of the chimeric mice marked a turning point in the stem cell story in two ways. In a scientific sense, Mintz and Illmensee had elegantly shown that teratoma cells can indeed support normal development, given certain still-unidentified signals. Teratoma cells, and the EC cells that resemble them, had instantly evolved from a peculiarity to a potential way to investigate early development. But in a bioethical sense, they had brought secretiveness and competition to science. "I remember how crushed my dad was," recalls Anne Wheeler, Leroy Stevens' daughter. "He had been working on this and had given Bea his mice, and told her all his ideas to do this. But she got there first. She had more money and a bigger lab. I remember my dad talking about how he saw the end of an era in the scientific community. Before that time, people couldn't wait to tell each other what they'd found, and to help others to get results. Later, scientists in the same lab would hide information from each other."

Once Mintz and Illmensee's chimeric mice had rescued teratoma cells from a cancer-only fate, many other researchers adopted the cells in studying early embryogenesis. But rather than repeatedly performing surgery to passage teratomas through countless mouse generations, attention turned to culturing EC cells. Still, EC cells weren't ideal for scrutinizing normal development because their strain 129 origin condemned at least some of them to the path toward a tumor. EC cells looked quite like normal cells, however, so researchers began to wonder whether truly normal early embryo cells, too, could be cultured and coaxed to recapitulate development.

In 1981, two research groups, one led by Martin Evans at the University of Cambridge and the other by Gail Martin at the University of California, San Francisco, working with mice, isolated normal embryo cells that closely resembled EC cells. These normal cells were part of the inner cell mass, a small grouping of cells that clings to the inside surface of the blastocyst, which is the hollow ball that precedes formation of the tri-layered embryo. The inner cell mass cells go on to form the embryo, whereas the other blastocyst cells form supportive membranes and structures.

Both research teams got the inner cell mass cells to grow in culture, providing mouse embryonic stem (ES) cells, rather than embryonal carcinoma (EC) cells. Brigid Hogan, a professor of cell biology at Vanderbilt University School of Medicine in Nashville, explains that "to make ES cells, you are asking a few cells in the blastocyst to continue proliferating. As long as they keep on multiplying and not differentiating, you can go on growing them." Adds John Gearhart, who was a postdoctoral researcher in Beatrice Mintz's lab at the time of the surprise announcement and is today a professor of gynecology and obstetrics at Johns Hopkins University School of Medicine, and whose group was one of the two that independently reported culturing human ES cells in 1998, "Evans and Martin took blastocysts from different mouse strains, and a certain percentage developed in long term culture into ES cells. From this procedure, all ES cell work has been done, in different labs." These 1981 papers are the ones that *Science* magazine cites as founding the field of stem cell research.

Several researchers soon discovered that the cultured mouse ES cells have a tendency, if left to their own means, of differentiating into seemingly random cell types. One investigator recalls as a graduate student glimpsing, from the corner of her eye, a tiny dish among many whose contents began to pulsate, reminiscent of the early work on embryoid bodies. The ES cells there had formed cardiac muscle tissue, which was doing what cardiac muscle tissue does—beat. However, ES cells in culture did not yield the entire repertoire of specialized cells, as an inner cell mass cell can as it gives rise to an embryo in normal uterine surroundings. It seems that something about life in a culture dish limits developmental potential, if only slightly. By adding various hormones and growth factors to see what would happen, researchers soon discovered that something called leukemia inhibitory factor keeps ES cells in culture in the undifferentiated state.

While Evans and Martin were perfecting culture of ES cells from mice, in 1982 researchers at the Wistar Institute in Philadelphia established cultures of human EC cells from testicular tumors. These cells were, of course, potentially cancerous and so not likely to have any medical use, but they seemed so similar to cells of the early embryo that researchers thought they could be useful in studying development. If injected into mice these human EC cells formed tumors, but when placed in dishes, like the mouse ES cells, they would spontaneously differentiate. It looked as if changing the culture conditions could either encourage specialization or keep the cells with a developmental clean slate.

Even though EC cells came from humans, they proved poor substitutes for mouse ES cells in the research lab in other ways besides their cancerous origin. The human EC cells just weren't robust. Many would simply die in culture, and very few would form embryoid bodies. "Human EC cells, which I worked with a bit, long ago, were dreadfully difficult to keep alive and healthy in culture," recalls Thomas Doetschman, a professor of molecular genetics at the University of Cincinnati College of Medicine. Plus, EC cells often were missing a chromosome or two, making them unlikely to develop much beyond a few cell divisions.

KNOCKOUTS BEGIN WITH EMBRYONIC STEM CELLS

Developmental biologists circa 1982 wanting to study early mammalian development had a choice: mouse ES cells or human EC cells. For the most part, the mouse cells won. Soon, the focus of research turned from using ES cells to probe early development to using them in a new technology. Gene targeting, soon to bear the more colorful name of knockout or knockin technology, was a marriage of sorts between stem cell biology and genetic engineering. It enabled researchers to create multicellular organisms with a specific genetic change in every cell—either an existing function obliterated ("knocked out") or a new one added ("knocked in"). These techniques are important to the story of stem cell science because they provide the context within which many biologists knew of ES cells. Gene targeting is also important because it raised bioethical concerns long before the rest of the world even knew what stem cells were.

The evolution of gene targeting illustrates how research threads become intertwined to knit a new technological fabric. Several observations and experimental approaches contributed to the birth of this technology. First, extension of Mintz and Illmensee's famous mouse chimera experiment showed that ES cells not only become parts of normal animals, but that those cells could become part of the germline, the cells destined to become sperm or egg. This meant that chimeras could be bred to each other, and in cases where both male and female had an ES cell–based germline, a certain percentage of the offspring would descend completely from the ES cells (Figure 6.5). This roundabout route of making chimeras was necessary because ES cells by themselves do not develop into normal embryos, just conglomerations of embryonic and adult tissues.

Great ideas tend to occur to several people at once, and this is what happened with the knockout mouse saga. Researchers realized that they could manipulate ES cells in the laboratory, adding or deleting genes and letting the cells grow in culture to see what kinds of tissues they could become. Then, they could mix ES cells with normal embryo cells, breed chimeras, mate them, and look for offspring that had the targeted change in every cell. Initial efforts used marker genes whose presence would be easy to detect at the cellular level, such as genes conferring sensitivity or resistance to a particular drug or toxin. Experiments also used coat color genes, so that different offspring classes could be easily distinguished from one another.

Making mice with genetic changes stitched into every cell was already possible as an outgrowth of recombinant DNA technology called transgenics. But the technique was imprecise. In transgenic technology, a researcher delivers a gene of interest into a fertilized egg, so that it is retained and perpetuated in all cells of the organism that develops. However, the transgene can insert into the genome anywhere, perhaps where it won't be expressed or where it might actually cause damage. Plus, the introduced gene has to compete with the effects of the counterpart gene already present. But performing genetic manipulations on ES cells, instead of fertilized eggs, adds predictability and precision because the engineered cells can be examined *before* they are used to generate mice.

In 1986, Tom Doetschman, then a young postdoctoral worker at the Max Planck Institute in Tubingen, Germany, added a gene to ES cells that enabled them to make an

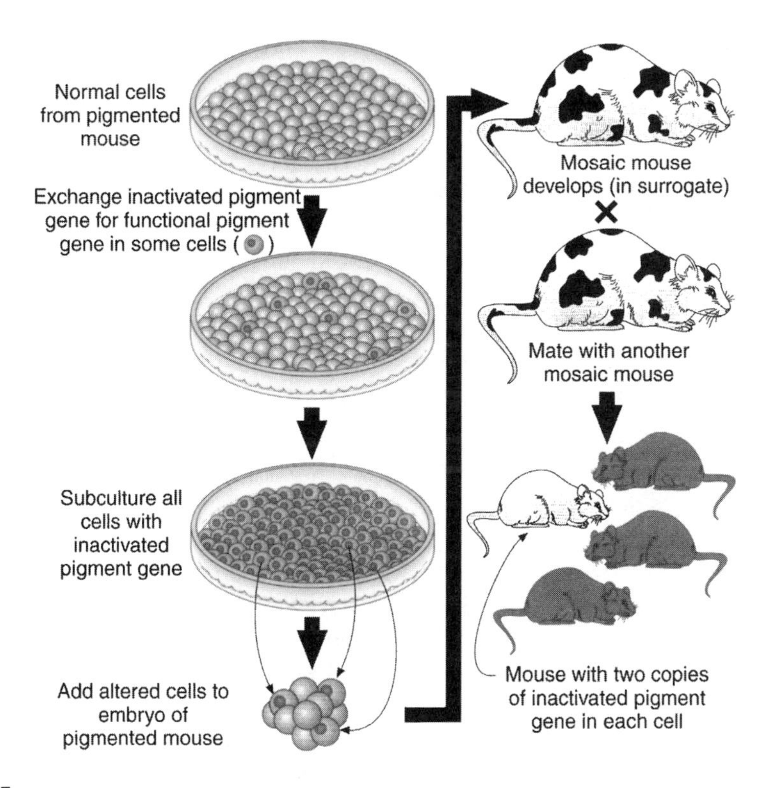

FIGURE 6.5

Complex breeding schemes using chimeric mice are necessary to derive offspring with the targeted gene in every cell. The process begins with ES cells.
Lewis R. Human genetics: concepts and applications. *3rd ed. Boston: McGraw-Hill, 1999:320. Reproduced with permission of The McGraw-Hill Companies.*

enzyme (neomycin phosphotransferase) that renders them resistant to a particular toxin. He made chimeric embryos that developed normally into mice, then bred them to each other and selected mice built entirely of the marked cells. After the results were reported in a prestigious scientific journal, Doetschman braced for the anticipated press fanfare at genetically engineering a mouse. But it didn't happen. "At first, people were very concerned. Can you do this with humans? What are the implications? But then the interest died away," he recalls.

It isn't clear why the idea of recombining genetic material in 1975 ignited a media frenzy, as did the birth of a cloned sheep 21 years later, yet the ability to so specifically genetically engineer a mouse in 1986 had no impact. Perhaps public awareness of scientific results follows its own development program, an ebb and flow that somehow failed to notice the birth of the knockout/knockin mouse that would so profoundly affect biomedical research in the years to come. Mice are, in fact, so genetically similar to humans that one prominent researcher says, laughingly but not entirely in jest, that a mouse cell could be altered to become a human cell.

Although the world at large failed to take notice, ES cell research proceeded full speed ahead. In 1987, Doetschman, then working in Oliver Smithies' lab at the University of Wisconsin, added the targeting part of gene targeting.

Two years earlier, Smithies had pioneered investigation of homologous recombination, which is the natural tendency of an isolated piece of DNA to locate its complement within an organism's chromosome and then switch places with it. A piece of DNA with the sequence ATCGCTAC, for example, would align itself with TAGCGATG wherever it appeared in a chromosome and then swap itself in. Adding a homologous recombination step to transgenic technology introduced greater precision. Instead of the gene of interest arriving and inserting at random, as it did in transgenesis, it would now nestle into its normal chromosomal home, jettisoning the existing gene. With homologous recombination, a gene of interest could *replace* the existing gene, not compete with it.

Doetschman used homologous recombination to knock in a functional copy of a gene that encodes an enzyme called HPRT into mice that lacked the gene. HPRT was commonly used for genetic studies because it is carried on the X chromosome and therefore a mutation need only affect one copy of the gene in males, with their lone X chromosome, to be detectable. And it was detectable because cells with mutant HPRT survive in the presence of a toxin that kills normal cells.

HPRT was also of interest because when it is missing or abnormal in humans, a horrendous disease, Lesch-Nyhan syndrome, results. Affected boys are mentally retarded and uncontrollably mutilate themselves, biting off their fingers and lips and chewing into their shoulders if unrestrained. Even though mouse and human are so alike genetically, for reasons unknown mice missing the enzyme seem unperturbed.

Over the next few years, several groups of researchers conducted many other variations on the HPRT gene-swapping theme. In 1989, Mario Capecchi's group at the University of Utah replaced a mouse's working copy of the HPRT gene with a nonfunctional HPRT gene, creating the first knockout mouse. Capecchi pointed out at the time that the advent of these mice meant that any gene could now be modified in any way, from any species from which functional ES cells could be obtained. Capecchi also suggested then, quite prophetically, that researchers could derive human stem cells from somatic tissues (rather than from embryos) and correct faulty genes *in vitro*, then deliver the cells therapeutically in implants.

Knockout mice quickly became a laboratory favorite, giving geneticists a way to vanquish a gene's function so that they could infer what it normally does. Surprisingly, many mice suffering a specific knockout seemed none the worse for the experience, revealing an unsuspected plasticity encoded in the genome. That is, destroy one gene, and a different one can take over the function, perhaps by rerouting metabolism. In 1993, at the request of the National Institutes of Health and many researchers, the Jackson Laboratory opened its Mutant Mouse Repository, which offers transgenic and knockout/knockin mice. Today, much of human genetic research is conducted first on these mice.

ENTER EMBRYONIC GERM CELLS

In the late 1980s, while gene targeting was catching on, ES cells were available only from mice and hamsters, and researchers tried doggedly to isolate and nurture the cells from pigs, cows, birds, and other animals. The work was slow and laborious, harder than it had been in mice. This was because in mice, the blastocyst that surrounds the precious inner cell mass collapses and breaks away in culture on its own, leaving behind clumps of ES cells. In other animals, especially primates, the blastocyst must be carefully removed or the inner cell mass dies. Was there a better way to get stem cells? Indeed there was. Leroy Stevens had found a source years earlier (Figure 6.6).

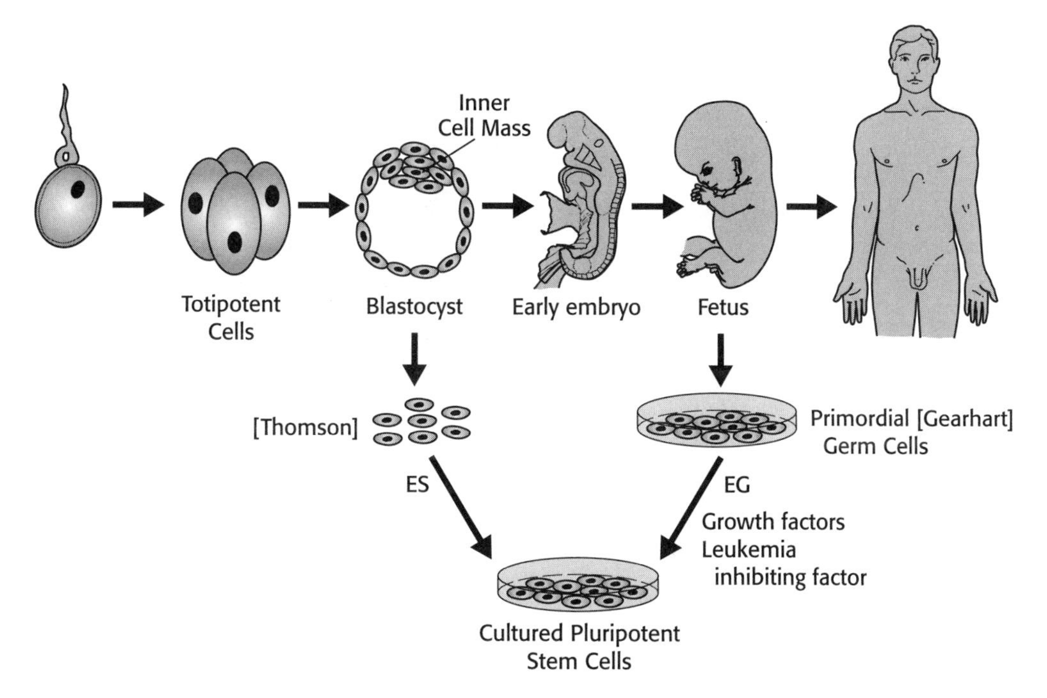

FIGURE 6.6

Sources of embryonic stem (ES) and embryonic germ (EG) cells. Inner cell mass cells yield ES cells. Primordial germ cells from a fetus yield EG cells. Both ES and EG cells are valuable because they are pluripotent, perhaps even totipotent. The inset shows human ES cells.
National Institutes of Health. Adapted.

Recall that Stevens had traced teratomas back to primordial germ cells arriving at the genital ridge 12 days after conception. But he could first detect these cells on day 7 because they produce an enzyme, alkaline phosphatase. Midway through the twelfth day, the cells migrate from an area in the hindgut next to the allantois (one of the membranes that supports the embryo) to the genital ridge. "During the migration, the cells divide rapidly, starting with 10, maybe 100 primordial germ cells, reaching up to 100,000 as they enter the developing gonad. Once there, they settle into an association with nurse cells that cluster around and feed and protect them. They stop dividing, and differentiate more extensively," explains Hogan. The cells remain in a state of suspended animation until the mouse-in-the-making starts manufacturing sperm or ova.

Primordial germ cells could be another source of ES cells because they demonstrate, when they form teratomas, that they can differentiate into a variety of cell types. "If you culture cells from this region of the embryo, you can derive long-term cultures with all the properties of ES cells," reports Gearhart. But that was easier said than done, for primordial germ cells didn't take well to existence in a lab dish. The cells would start out vigorously and then cease dividing after about a week in the lab. They were actually acting quite naturally, slowing down the cell division rate as they would in an animal. "The trick is to catch these cells when they are still proliferating, and there are receptors on the cell surfaces for various growth factors. You have to catch them as they migrate, or just after they have arrived at the gonad. Then if you culture them, they may continue proliferating. They revert in culture to an earlier embryonic state, and behave very much like cells set aside as the inner cell mass," explains Hogan.

In 1992, Hogan's group and Peter Donovan's team at the National Cancer Institute reported conditions necessary to keep primordial germ cells happy in cell culture. The recipe included adding fibroblast growth factor and leukemia inhibiting factor. "To keep them proliferating, we add a cocktail of growth factors, and they behave as ES cells," says Hogan. The cells became known as embryonic germ (EG) cells. They are also called ES-like cells because of their similar repertoire of activities. Like ES cells, EG cells form the following:

- Embryoid bodies in mouse abdomens
- Differentiated cells in teratomas when placed in immunosuppressed mice
- Specialized cells in culture if left alone
- Parts of mouse chimeras

In addition, EG cells are much easier to harvest from animals than ES cells are. They need only be scooped out of ovaries rather than painstakingly retrieved from delicate, tiny blastocysts.

BEYOND THE MOUSE

Isolating human ES cells was a logical next step once researchers perfected the ability to gently liberate inner cell mass cells in culture from the blastocyst. But a few species preceded *Homo sapiens.*

Tom Doetschman accomplished the feat in 1988 for hamsters, making gene targeting possible for this rodent that is a closer model for lipid metabolism and cardiac function in humans than are mice. By the mid-1990s, ES cell technology transcended the realm of the rodent (Table 6.1). Many recognize the magic touch of James Thomson of the Wisconsin Regional Primate Research Center in teasing the coveted inner cell mass cells from primate blastocysts in culture as a giant step forward. Research followed a now-familiar series of steps:

TABLE 6.1

Embryonic Stem Cells Go Beyond the Mouse

Species	Research Group	Year	Topics Studied	Source
Hamster	Tom Doetschman (Univ. of Wisconsin)	1988	Lipid metabolism, cardiac defects	icm
Rhesus monkey	James Thomson (Univ. of Wisconsin)	1995	Early embryonic development	icm
Marmoset	James Thomson	1996	Reproduction	icm
Chicken	B. Pain (Lab. of Molecular Biology, Lyon, France)	1996	Agricultural traits	icm
Pig	Gary Anderson (National Cancer Institute)	1997	Cardiovascular research, xenotransplantation, agricultural traits	EG

*EG = embryonic germ cell; icm = inner cell mass from blastocyst.

Cells would first be cultured indefinitely, then shown to be able to become parts of chimeras with normal cells, and then, the coup de grace, shown to be able to form germline tissue in chimeras so that researchers could breed animals made entirely of stem cells.

Having ES cells and the ability to develop animals consisting solely of them opened up new types of animal models of human disease. For example, ES cell–derived rhesus monkeys provide a valuable stand-in for studying such early developmental events as activation of the offspring's genome, the structure and function of the placenta, and the form of the early embryo. But sometimes nature interferes with technology. The mating behavior of the rhesus monkey presented obstacles to reproductive biologists seeking many monkeys. Each female has only one offspring, she doesn't become sexually mature until she is 4 or 5 years old, and her menstrual cycles do not respond well to the hormone treatments necessary to synchronize them so that researchers can collect embryos. One way around the inconvenient reproductive cycle of rhesus monkeys is to use digestive enzymes to split early embryos into separate cells, and then allow the cells to develop into individual animals in surrogate mothers. But Thomson found another approach—a more cooperative primate, the marmoset. It is sexually mature at 18 months, has twins or triplets, and its ovarian cycle favors easy collection of blastocysts.

Meanwhile, work was intensifying on obtaining human ES cells. Two groups led the way. Thomson's team at Wisconsin used the inner cell mass–blastocyst approach; John Gearhart, at Johns Hopkins, took the EG cell route.

FINALLY—HUMAN EMBRYONIC STEM CELLS

Perhaps it was the midsummer announcement. Perhaps it was an overdose of Dolly the cloned sheep. Whatever the reason, when the Johns Hopkins University School of Medicine sent out a news release dated July 28, 1997, announcing that John Gearhart and postdoctoral fellow Michael Shamblott had cultured human ES cells for seven months, hardly anyone in the media took note. It's not as if the news release wasn't exciting. It began, "Researchers at Johns Hopkins have developed the first human embryonic stem cell lines, cells that theoretically can form all the different cells and tissues of the body." But only those familiar with mouse knockouts might have become at least intrigued, if not alarmed, at the possibilities that the research suggested.

Reporters from the mainstream media were not exactly beating down Gearhart's door. When I called him to arrange an interview to report on the work for *The Scientist,* a bimonthly newspaper for life scientists, I was shocked when he answered his own phone and chatted away for an hour! And so *The Scientist* headlined the article "Embryonic Stem Cells Debut Amid Little Media Attention."

For a while, nothing happened in response to Gearhart's work. But the lack of attention might have reflected how science generally proceeds: New results are published in a peer-reviewed journal, and perhaps presented before that publication at a meeting. Gearhart had discussed his lab's efforts at an ethics workshop at the 13th International Congress of Developmental Biology in Snowbird, Utah, the week of July 12, which had precipitated the news release. "I was invited to attend an ethics workshop at the congress to discuss issues surrounding human cloning and human ES cells, and thought long and hard about what to say, because our work was not yet complete. But that was a good opportunity to get a discussion going on human stem cell work, and to tell people that it was coming sooner rather than later. So we presented our work." He and Shamblott had indeed maintained human ES cells in culture for months, but they had yet to show that the cells could differentiate into ectoderm, mesoderm, and endoderm. At the time, Thomson predicted that "the press may be holding back because [the work] hasn't been published yet. When Gearhart demonstrates the three lineages [of specialized cells] developing from the cells, the press will pick up on it." He was right.

When Gearhart reported the experiments showing that his ES-like cells differentiated *in vitro* into derivatives of all three embryo layers in the November 1998 issue of *Developmental Biology,* and Thomson reported similar experiments in the November 6 issue of *Science,* the media finally noticed. Both research groups had transplanted their cells to immunodeficient mice to see what they would become, and cataloged a variety of tissues representing all three germ layers. And both pointed out potential applications in basic research, drug discovery, and transplantation medicine.

Developmental biologists hailed the work. "Human ES cells would be very useful to understand the steps in differentiation. We could show, for example, which genes are turned on and which growth factors are necessary for cell specialization," explains Hogan. ES cells stimulated to differentiate into human tissues or organs would better approximate pathogenesis than mice endowed with a few human genes. "There are a lot of mouse models of human disease, but there is nothing like working with truly appropriate material," notes Gearhart. And new drugs that might harm an embryo or fetus could finally be tested, instead of waiting for medical disasters to occur.

It was obvious that ES cells could be used to generate replacement parts if researchers could figure out the precise recipes for urging them into specific cell fates. One approach, called therapeutic cloning, would transfer the nucleus from a person's somatic cell into an enucleated egg to generate a blastocyst corresponding to that person's particular genome (Figure 6.7). Then the inner cell mass cells would be used to derive ES cells that would give rise to the cell type necessary to correct a particular problem, such as providing muscle tissue. It is cloning, but only for a few days. However, the creation of a new individual, albeit just a ball of cells, raises serious ethical concerns.

A more ethically acceptable strategy is to maintain cultures of ES cells or their derivatives directly, skipping an embryo stage. Then the cells would receive the appropriate biochemical signals to switch on a particular developmental pathway, providing material for cell implants that could be used to augment or replace diseased tissue. "Our work on the mouse suggests that if we had bucketloads of human ES cells, we could indeed turn them into cells that we could implant into patients with multiple sclerosis," explains Ronald McKay, chief of the laboratory of molecular biology at the National Institute of Neurological Disorders and Stroke.

ENTER ETHICS AND POLITICS

Many people, perhaps sensitized by the hullabaloo over Dolly, envisioned warehouses of human embryos being harvested for spare parts. The sources of Gearhart's and Thomson's cells attracted attention, although their experiments were approved by institutional review boards. The blastocysts that yielded their inner cell masses to Thomson's lab were "leftovers" from *in vitro* fertilization clinics, used with informed consent from the donors. Using private university funds, Gearhart acquired primordial germ cells from 5- to 7-week-old embryos aborted at a clinic. None of the material came into existence to fulfill the requirements of a research protocol.

Arthur Caplan, director of the Center for Bioethics at the University of Pennsylvania, points out that the blastocysts and embryos would have existed whether or not they were

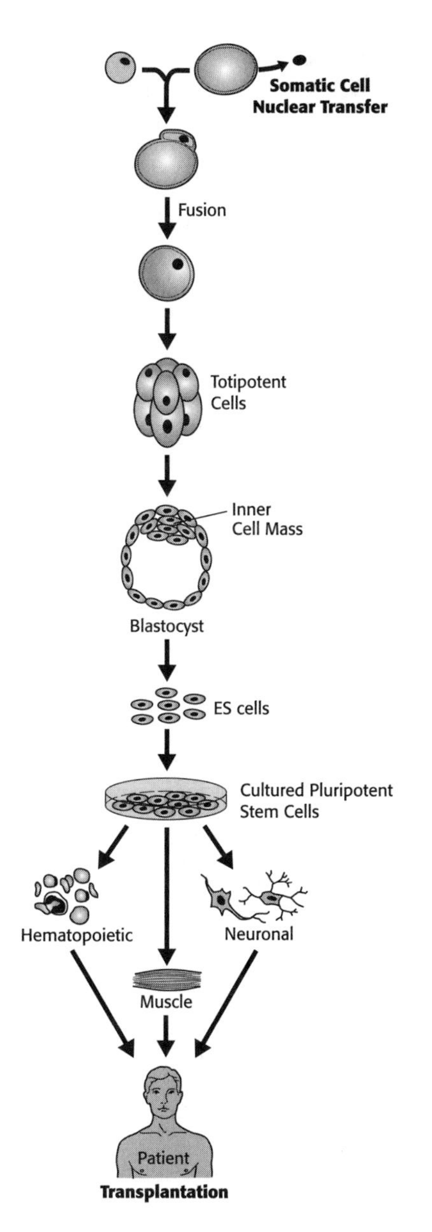

FIGURE 6.7

Therapeutic cloning creates a short-lived clone of an individual to obtain its inner cell mass cells, which can be coaxed in laboratory culture to become embryonic stem cells. These ES cells can then be used to generate needed tissue. This approach is not yet being done, and may never be done due to ethical objections to the intentional use of an early embryo to provide spare parts.
Adapted with permission from Solter D and Gearhart J. Putting stem cells to work. Science *1999;283:1469. Copyright 1999 American Association for the Advancement of Science.*

used to generate ES cells. Ronald McKay raises a different argument, one already evoked to justify the fetal cell implant research that has achieved some success in treating people with Parkinson's disease for whom drugs no longer are effective. As with Parkinson's disease, McKay argues that a few initial embryos or fetuses could generate material that could provide treatments for thousands of people. Gearhart agrees. "The cells are a renewable resource. We could bank cells, without having to use additional abortion material."

A big fear stemming from the prospect of ES cell technology is whether it would ever be used to clone knockout or knockin people. Although this is theoretically possible, scientists uniformly say they will not create human clones, from ES cells or otherwise. Anticipating such questions of germline manipulation and cloning, Gearhart said in 1997, "[W]e will not perform any experiments aimed at genetically engineering the human germline in my lab or anywhere at Hopkins—it is not ethically acceptable. That's one of the reasons we chose to present our studies in an ethics forum. We wanted to begin the process of establishing a set of guidelines for the ethical use of cells of this type." However, ES cell technology greatly speeds the cloning of experimental or agricultural organisms by eliminating the breeding-of-chimeras stage, first done with mice in late 1999.

Back in 1998, with the stem cell announcements arising seemingly from nowhere, lawmakers were caught by surprise between the public reaction against any experiments involving human embryos, and a research community anxious to explain just what such work would involve and what it might promise in terms of medical benefits. And so politicians turned to their advisors, struggling once again to quickly compensate for the fact that Biology 101 was, for most, a distant memory.

In October 1998, Congress reiterated a 1994 ruling preventing use of government funds for "the creation of a human embryo" for research, or "research in which a human embryo or embryos are destroyed, discarded or knowingly subjected to risk of injury or death." Because abortion is legal, and during an abortion an embryo is definitely "knowingly subjected to risk of injury or death," the restriction seemed to focus on the issue of intent to use the material for a particular purpose. The law was valuable, however, because it clearly distinguished an embryo from a human subject or fetus, already protected by law. And the legal definition was both precise and inclusive, calling an embryo a structure "that is derived by fertilization, parthenogenesis, cloning, or any other means from one or more human gametes or diploid cells." That about covered all the bases except for the stork or extraterrestrial intervention. Private or industrial funds could still be used to support research on human embryos.

But the issue was far from settled. In November 1998, right after the Gearhart and Thomson papers appeared, President Clinton charged the National Bioethics Advisory Committee to review the field and make specific recommendations for how federal funds could be used.

Meanwhile, as a backdrop to the ethical objections, the scientific front was abuzz with excited talk about the potential benefits of therapies using stem cells from embryos, fetuses, and adults. President Clinton listened. He emerged momentarily from his personal woes to reconsider and backpedaled a bit, saying that perhaps the law had been written too hastily back in 1994. So in January 1999, Harold Varmus, then director of the National Institutes of Health, placed a moratorium on the ban, allowing research on human pluripotential stem cells—which includes those from embryos. But before you could say "inner cell mass," 70 members of the House of Representatives sent a letter to the Department of Health and Human Services in February 1999 requesting rescission of the temporary ruling allowing federal funds for research on embryos to obtain ES cells. They sought to ban all research on stem cells obtained from human embryos and fetuses.

The scientists hurled the next volley, with a letter in the March 19 issue of *Science*, signed by a who's who of biologists, that listed the many potential benefits of the technology. They wrote:

> In addition to helping to unravel processes underlying cell differentiation and biological development, the use of human pluripotent stem cells could potentially reduce the number of animal studies that clinical trials require for drug development and testing . . . . Stem cells could be used to generate a long list of cells and tissues that could be used for transplantation.

A description of specific applications followed. The conclusion was clear—banning use of federal funds for stem cell research would slow development of these medical technologies by forcing researchers to seek funding from outside the government.

The winds began to shift in favor of stem cell research when the National Bioethics Advisory Committee notified the press in July 1999 that it had deemed the work "compelling and worthy of pursuit" and would charge the National Institutes of Health with developing guidelines for research and an oversight system to monitor procurement of stem cells. The committee gave its draft guidelines to President Clinton on September 13. He announced the next day, "The scientific results that have emerged in just the past few months already strengthen the basis for my hope that one day, stem cells will be used to replace cardiac muscle cells for people with heart disease, nerve cells for hundreds of thousands of Parkinsons' patients, or insulin-producing cells for children who suffer from diabetes."

The report spelled out specific conditions and limitations:

- Government funds cannot be used to create embryos for research purposes, but such funds can be used to work on material obtained with other resources. Translation: Get the cells privately.
- Donors of materials, such as couples with extra embryos conceived for *in vitro* fertilization, must be informed of potential use of their cells or embryos for research. People are not to be coerced or offered incentives to donate material, and such material cannot be identifiable to a particular individual.
- A "Human Pluripotential Stem Cell Review Group" will oversee compliance with the regulations and will approve all research protocols that use federal funds. This will provide some control over the private procurement of inner cell mass and primordial germ cells.

The American Society of Human Genetics supported the proposed guidelines, but added a scenario that bioethicists hadn't considered—intentional use of abnormal cells. "To understand mechanisms of disease, researchers in human genetics often use cells or cell lines derived from individuals who are genetically or chromosomally abnormal. Stem cell lines derived from such abnormal embryos will be extremely valuable to study disease mechanisms," pointed out Uta Francke, a professor of genetics at Stanford University, representing the board of directors of the 6000-member genetics organization. She requested that cell samples be labeled to indicate the abnormality, but not the donor identity. Abnormal cells might come from *in vitro* fertilization donations discarded because of extra chromosomes or from preimplantation diagnosis, in which couples with known high risk for a particular disorder have their eight-celled embryos tested, allowing only healthy ones to complete development.

When John Gearhart spoke at the developmental biology conference in 1997, he anticipated the negative public reaction to his use of human embryos, as did others. Said Ronald McKay at the time, "The idea of harvesting embryonic tissue in an uncontrolled way may be distasteful. I prefer the idea that we could grow cells in the lab with certain useful, highly differentiated properties." Although nurturing tissues in a laboratory from adult stem cells is a bioethically more pleasing scenario than using embryo cells, at the present time adult stem cell technology is just not as promising. "There are things that stem cells taken from adults won't and can't do, and many can talk the talk but not walk the walk. We need to wait to determine the role for ES cells in the technology. I would hate to see the research not supported until we can see which types of stem cells work in which tissues," cautions Gearhart today. Table 6.2 describes some of the experiments already done with ES cells in mice, and their potential medical applications. Gearhart's lab is currently testing the abilities of a variety of human stem cells to correct genetic defects in mouse models of human diseases.

ADULT STEM CELLS: BONE MARROW TO BRAIN, LIVER, MUSCLE, AND BEYOND

Using stem cells that persist in an adult animal's body is not a new idea. Researchers for decades have been scrutinizing the many cells that compose bone marrow, meticulously tracing cell fates and seeking the chemical cues that guide cells through the diverging pathways of differentiation. The medical value of controlling the process is obvious.

TABLE 6.2

Mice Now, Humans Next?

In Mice, ES Cells Become	Medical Technology Application
Hematopoietic cells (red blood cells, white blood cells, mast cells, and macrophages)	Bone marrow transplants, blood substitutes
Blood vessels with blood cells	Coronary bypass grafts
Skeletal muscle that contracts	Implants to treat muscular dystrophies
Heart muscle that contracts	Implants to treat muscular dystrophies and cardiovascular disease
Neurons with action potentials, glia	Implants to treat neurodegenerative disorders, implants that restore some mobility to rats with damaged spinal cords
Precursor pancreatic cells	Implants to treat diabetes
Bone marrow stem cells	Implants to replace or augment liver function

For years bone marrow stem cells have been used to treat blood conditions and immune deficiencies and to reconstitute a hematopoietic system ravaged by chemotherapy, enabling cancer patients to withstand higher, more effective doses of the toxic treatments. Today many new parents store umbilical cord stem cells from their offspring, a health insurance of sorts that might one day provide certain nonrejectable replacement parts. One researcher, aware that stem cells are probably present in tissues other than bone marrow, although they haven't yet been found there, sampled and froze various tissues from his infant son.

The Holy Grail of adult stem cell research is a bone marrow component that goes by two names. The most common term is *marrow stromal cells,* because they seem to arise from the supportive, or stromal, cells of bone marrow. These cells are also called *mesenchymal stem cells* because they resemble mesenchyme, a tissue in an adult that can give rise to many cell types. Mesenchyme is the body's storehouse of potential spare parts—pockets of unspecialized cells, in a variety of places, that can migrate to an injury and, responding to signals in the environment, embark on a normal developmental pathway to become what's needed. The marrow's stromal cells are likely their version of mesenchyme.

The developmental potential of marrow stromal cells is more restricted than that of ES cells, but what they can become is still extraordinarily useful. These cells can differentiate into connective tissue and its derivatives blood, bone, and cartilage, as well as fat cells, muscle cells, nerve cells, and the outer layers of blood vessels. Nineteenth-century pathologist Julius Cohnheim's experiments first revealed that bone marrow sends cells to wound sites, where they differentiate into healing fibroblasts. In the 1970s, Alexander Friedenstein, at the Gamaleya Institute for Epidemiology and Microbiology in Moscow, continued Cohnheim's work by plating bone marrow onto culture dishes and noticing that only some of the cells stuck. These adherent cells aggregated into groups of two, three, or four, and they became increasingly spindle-shaped when they divided in culture over several days. Eventually, the cells gave rise to bone and cartilage. By the 1980s, Friedenstein and others had succeeded in getting crude extracts of adherent bone marrow cells to differentiate into bone, cartilage, fat, and muscle.

In the 1990s marrow stromal cell research focused on manipulating culture conditions to route development toward specific fates. For example, in marrow extracts from mice, rats, rabbits, and humans, adding vitamin D and certain cytokines and hormones directs the cells to differentiate as bone, cartilage, or fat cells, depending on the particular recipe. Adding a different cocktail that includes the drugs 5-azacytidine and amphotericin B transforms marrow cells into muscle cells that spontaneously aggregate into tiny pulsating tubules.

Bone marrow is a very complex tissue, and researchers have long wondered whether a single stromal cell can yield several cell types or whether a mixture of cell types is responsible for the various developmental fates. Two strategies can be used to follow the fates of a single, pluripotent cell: identify its descendants and show that they are identical, or watch a single progenitor cell divide and give rise to a colony, or clone (Color Plate 4).

Ranieri Cancedda, at the Center for Biotechnology in Genova, Italy, took the first approach, isolating subpopulations of bone marrow from 200 volunteers. He inferred clonality by demonstrating that the cells have the same cell surface features and patterns of gene expression. Once satisfied that the cells in an individual were probably descended from a single progenitor, Cancedda's group observed the developmental trajectories of cells taken from the same clone. "Each clone has been expanded in culture and analyzed for its capacity to differentiate under the appropriate culture conditions in three lineages—adipogenic [fat], chondrogenic [cartilage], and osteogenic [bone]. When the cells were expanded in the presence of fibroblast growth factor, about 33 percent of the clones were tripotential," Cancedda explains. In fact, cells were more likely to have two or three fates than a single one. This pluripotentiality, coupled with the consistent patterns of gene expression, provides powerful evidence that these groups of differentiating cells indeed descend from single stem cells.

Taking the more direct but time-consuming approach, researchers at Osiris Therapeutics separately placed individual marrow stromal cells from three donors, plus single cells from two different fibroblast cell lines, in cocktails of growth factors, hormones, vitamins, cytokines, and serum known to stimulate differentiation as fat, cartilage, or bone. Within 3 weeks, the telltale signs of cell specialization were unmistakable in the cultures derived from the donors' cells: Future adipocytes ballooned with fat, the cartilage-to-be churned out characteristic proteoglycans and type II collagen, and bone cells stockpiled calcium and alkaline phosphatase. In contrast, the adult fibroblasts remained, stubbornly, adult fibroblasts, an expected result because they are not stem cells. To make sure that what they were seeing was truly what was happening, the researchers also monitored which genes were expressed in each experiment, confirming that fat was fat, cartilage was cartilage, and bone was bone. "These cells are uniform in phenotype, derived from single cells, and when exposed to different culture conditions become all osteoblasts, which become osteocytes and lay down bone mineral. Or they become cartilage and lay down cartilage extracellular matrix and then become morphologically chondrocytes. Or, they become fat cells. It is very striking," says Osiris's Daniel Marshak.

As many laboratories confirm the long-suspected hypothesis that bone marrow contains rare cells that have broad developmental potentials, thoughts of medical applications of stem cells aren't far behind. Therapeutic products will likely be the stem cells themselves, delivered as implants to correct localized disease. "Hematologists have been doing this for years, with hematopoietic stem cells. They don't make blood in the lab and infuse it; they put in the stem cells. We're taking the same approach," says Marshak. Replacing blood is a straightforward application—others are more complicated. Consider how stem cell therapy might be used to treat a heart attack. "Current therapy provides new vasculature, using angiogenesis factors or cardiac bypass surgery. But neither of these approaches addresses the problem of adding new heart muscle cells. We are putting human mesenchymal stem cells into damaged heart muscle tissue to repopulate normal heart muscle. It will complement angiogenesis—we will be able to both restore blood vessels and replace muscle," Marshak says.

Stem cell therapy to provide connective tissues might even transcend the body's natural healing abilities. Cartilage, for example, does not heal well because it is a bloodless tissue, inaccessible to the growth factors and stem cells necessary for repair. Perhaps an outside source of stem cells can infiltrate the affected tissue. Similarly, osteogenic stem cells might be able to fill in bone removed surgically, a very unnatural sort of trauma. "If a person has osteosarcoma and a surgeon removes a substantial part of the femur, then that leaves a large gap that would not normally heal. Stem cells for bone regeneration will fill segmental gaps and large dental defects where you need a large chunk of bone and normal repair is not sufficient," explains Marshak.

Even though bone marrow is the best-studied source of adult stem cells, the eclectic tissue still holds surprises, revealed unexpectedly in recipients of bone marrow transplants. When recipients and donors are of opposite sexes, descendants of the donated cells in the recipient's body can be distinguished by their different sex chromosome constitution, with the help of fluorescent tags that highlight the Y chromosome found only in males. A woman recipient of a bone marrow transplant from a man, for example, eventually produces "male" blood cells. But in 1993, researchers at Emory University in Atlanta were astonished to find donated bone marrow descendants in the brains of five female recipients!

In 1999 came an even more unexpected bone marrow relocation—the liver. Evidence came from animal experiments inspired by the cross-sex human bone marrow transplants. Researchers transplanted marrow from male to female rats, then administered a drug that damages the liver and prevents it from regenerating. Histologists had long known that after such treatment, "oval cells" appear and then divide and differentiate to replace liver cells, but no one knew where these cells came from. The fluorescent tags gave the answer: The oval cells in the female rats' livers had the telltale Y chromosomes. "Our animal experiments using cross-sex bone marrow transplants clearly show that bone marrow-derived cells eventually become fully functional liver cells, quite probably through an intermediate oval cell. The next step is finding the bone marrow stem cell giving rise to the oval cell, or discovering the signal that the liver broadcasts to recruit such cells to the scene of injury," explains Bryon Petersen, the pathologist at the University of Pittsburgh School of Medicine who led the rat study. A few months later, a team from the New York University School of Medicine and the Yale University School of Medicine showed a similar marrow-to-liver healing path in mice. These findings suggest exciting applications. "If we can isolate these cells, we have targets for gene therapy and the possibility of stem cell transplantation, rather than whole-organ transplantation. There's also the possibility of expanding cells in culture and creating an artificial liver," says Neil Theise, of New York University.

Bone marrow cells also have a relationship of sorts with muscle cells: The stem cells from one are able to replace tissue in the other. Muscle stem cells in the embryo are called myoblasts; in adult tissue, they are known as satellite cells, so called because they hug the muscle fibers.

In 1999, Louis Kunkel and Richard Mulligan and their colleagues at Harvard Medical School infused muscle stem cells into the bone marrow compartment of mice whose bone marrow had been destroyed. The mice also had a mouse version of the human disorder Duchenne muscular dystrophy—their muscle cells were unable to manufacture the muscle protein dystrophin. The infused muscle stem cells not only repopulated the vacated marrow cavity, but some of them also migrated to the muscles, where their descendants differentiated into muscle cells that could produce dystrophin! The treated mice recovered some muscle function. The ability to heal dystrophic muscles via the bone marrow or blood is extremely exciting because gene therapy efforts on boys with muscular dystrophy so far have failed. The human experiments introduced functional myoblasts directly into selected muscles, but provided only localized correction. This most serious form of muscular dystrophy affects the entire musculature. Peggy Goodell and her colleagues at the Baylor College of Medicine showed that muscle can also substitute for bone marrow. When they placed satellite cells into mice whose bone marrow had been destroyed, the animals produced blood cells of all types, with surface markers indicating that they had descended from the implanted muscle cells.

BRAIN MARROW REVISITED

Stem cell science is in its infancy, for we do not yet know where the body stores all its self-renewing cells. Their presence in bone marrow seemed obvious, but other body parts deemed too specialized to have stem cells are turning out not to lack them after all. Consider the brain.

For many years, the sheer number of neurons composing an animal's brain—billions—seemed to explain the ability to learn and to recover from injury. The raw material was there for new neural connections to form. The connections were new, but not the neurons, because these cells were thought to have reached the end of their dividing days. And so arose the admonition to limit use of mind-altering substances, lest one deplete a finite and ever-dwindling supply of brain cells. But a few lines of evidence suggested that among the nondividers were some cells in the brain that remained pluripotent.

In 1965, Gopal Das and Joseph Altman at the Massachusetts Institute of Technology reported that new neurons form in a rat's hippocampus, the memory center, confirming reports from as long ago as 1912 that rodent brains have some dividing cells. But the cells were scant, and were thought to arise, somehow, from existing neurons.

Then in the mid-1980s, Fernando Nottebohm at Rockefeller University contributed an important piece to the puzzle of brain renewal—he investigated how canaries and chickadees learn songs. He showed that new cells form in the hippocampus and other forebrain areas, and their number waxes and wanes with the seasons, peaking when birds must master their songs to communicate and survive. Clever experiments challenged the birds' song acquisition skills by placing their food supplies farther apart. The number of new hippocampal cells increased in parallel to the animals' increased need to sing to communicate food locations.

The discovery of new brain cells behind birdsong learning galvanized the search for such cells in mammals, but the investigation was difficult. "There was a technical problem in isolating and studying these cells. It's much easier to get to bone marrow than brain. In tissue culture we'd see some cells, but we didn't know what they were because there weren't any immunohistochemical markers to identify stem cells in the brain.

There were no surface markers as there are for bone marrow that would allow a researcher to say, 'this is a stem cell,'" recalls Dennis Steindler.

To search for elusive new neural stem cells in brains, investigators turned to bromodeoxyuridine (BrdU), an agent that marks dividing cells. By staining slices of brain, researchers discovered new cells in the hippocampus of tree shrews in 1997, and in marmosets in 1998—a few evolutionary steps closer to humans. Then a chance visit with cancer patients in 1998 led to finding these cells in human brains. Peter Eriksson, on sabbatical at the Salk Institute for Biological Studies in La Jolla in the laboratory of Fred Gage, learned of a clinical protocol in which BrdU was being used to track tumor growth in a group of patients with cancers of the larynx or tongue. Might BrdU also mark dividing cells in the hippocampus? His idea was rather grisly—asking the patients for permission to examine their brains after death to see if BrdU wound up in their brains, highlighting dividing cells. The people were willing.

Between 1996 and 1998, Eriksson collected five brains, and all showed apparently normal dividing cells in a part of the hippocampus called the dentate gyrus. Further experiments showed that a group of stem cells here are quite active, continually dividing to yield more stem cells, some of which quickly die and others that migrate and eventually differentiate into neurons. Work in rats revealed another outcropping of stem cells lining the ventricles. Some of these cells migrate to the olfactory bulb, the seat of the sense of smell. Here, depending on the precise mix of growth factors, the cells specialize as neurons, or as astrocytes or oligodendrocytes, which are types of glial cells that support, nourish, and communicate with neurons.

Now that neural stem cells had been identified in human brains, researchers could take advantage of a discovery by McKay's group back in 1990, who found that neural stem cells in rodents produce a cytoskeletal protein called nestin. "In the developing central nervous system, stem cells beautifully express this protein. When the cells differentiate, they lose nestin and express other proteins characteristic of different cell types," explains McKay. "It was a real breakthrough to be able to label stem cells in the developing nervous system and apply this to cells in culture too," adds Steindler.

The central nervous system stem cells demonstrate three localized fates. Expose them to leukemia inhibiting factor, ciliary neurotrophic factor, and various growth factors and they become astrocytes. Add thyroid hormone, and the cells become oligodendrocytes. But like the ability of bone marrow cells to differentiate as liver cells, neural stem cells can become more than just the obvious neurons and glia—they can form blood. In 1998, Angelo Vescovi's group at the National Neurological Institute Carlo Besta in Milan, Italy, showed that neural stem cells can produce blood cells in mice whose bone marrow has been destroyed.

This developmental plasticity had been noted 16 years earlier, by Perry Bartlett at the Walter and Eliza Hall Institute of Medical Research at Royal Melbourne Hospital in Victoria, Australia. Back in 1982, Bartlett had created suspensions of cells taken from mouse brains, then watched what the cells would become when injected into spleens of irradiated hosts and allowed to divide to form "colony forming units," or CFUs. He noted that for every 100,000 or so brain cells, on average 14 formed hematopoietic colonies. He suggested two explanations for finding blood-forming stem cells in mouse brains. Perhaps the blood transported the stem cells to the brain, where they might form a potential reserve supply of phagocytic cells. This was unlikely, he argued, because such stem cells are nearly absent in adult blood, and the blood-brain barrier blocks entry of cells into the brain. But the alternative hypothesis seemed, at this time when cells were not thought to transgress the primary germ layers set down in the embryo, even more implausible: "The other more highly speculative origin of CFUs is that they are derived from a stem cell that gives rise to both neural cell types and hematopoietic cells," Bartlett wrote in the *Proceedings of the National Academy of Sciences.* Yet the part of Vescovi's paper citing Bartlett's paper dismisses his findings as being due to "contamination."

Bartlett was probably ahead of his time, making a suggestion that seemed out of context to what was known, just as Carl Woese's discovery of a third form of life (see Chapter 3) and Stanley Prusiner's idea that a protein alone could be infectious (see Chapter 4) seemed renegade for many years. Scientists, however, report what they see, even if it isn't always logical, and they recognize that the unlikely *can* happen. But human nature being what it is—skeptical—and scientists being among the most skeptical of humans, generally the unexpected has to happen a few times before people start to pay attention. With mounting evidence, it appears now that stem cells can in fact alter their allegiances in response to certain, as yet not fully understood, biochemical cues. Clearly, we still have much to learn.

MORE QUESTIONS THAN ANSWERS

Stem cell biology is turning out to be another area of life science that is teaching us that we know far less than we had imagined. In particular, differentiation is not merely a sequential shutting off of certain genes or an assignment received very early in development that depends on which of three layers a cell finds itself in. Cell specialization is also a series of developmental decisions that is highly attuned to a very specific milieu of biochemical signals. McKay compares the situation to acquisition of languages. "One way to look at it is not as a series of highly discrete binary decisions, but as a response of a multipotential cell to different signaling environments. It is more like a sociology. People are the same no matter where they are born, but they rapidly become French if they are born in France, Spanish if born in Spain. We have to understand the language that a stem cell reads." Comprehending these exogenous controls on cell specialization will, ultimately, provide the information needed to engineer cell implants or to prompt organs to directly regenerate lost parts by tapping their own built-in stem cell supplies (Figure 6.8).

Stem cell biology is perhaps the most powerful life science to arise in a century, and so *Science* magazine's bestowal of the title of "breakthrough of the year" for 1999 to this field is apt. The field is providing researchers with long-sought peeks into exactly how cells communicate and respond to their environments in ways that enable them to build tissues. "And the best is yet to come. Stem cell biology will lead to amazing things in our understanding of cell reproduction and differentiation, tumorigenesis, and neurodegenerative diseases," says Steindler.

Despite the accolades and headlines, the ethics committees and the deluge of recent experimental results, researchers are confronting new questions about stem cells at a much faster rate than they can investigate the long-standing ones. At the same time, the quest to understand stem cells is uniting researchers from various fields, as evidenced at one recent conference. "One person who had investigated liver regeneration for a long time was talking, and I realized I hadn't paid attention to the work in the past because I was interested in bone marrow stem cells. Then someone spoke about muscles. We are all coming together to talk and to compare markers to sort out the similarities and differences among the stem cell populations in adult tissues. All of the stem cell people are

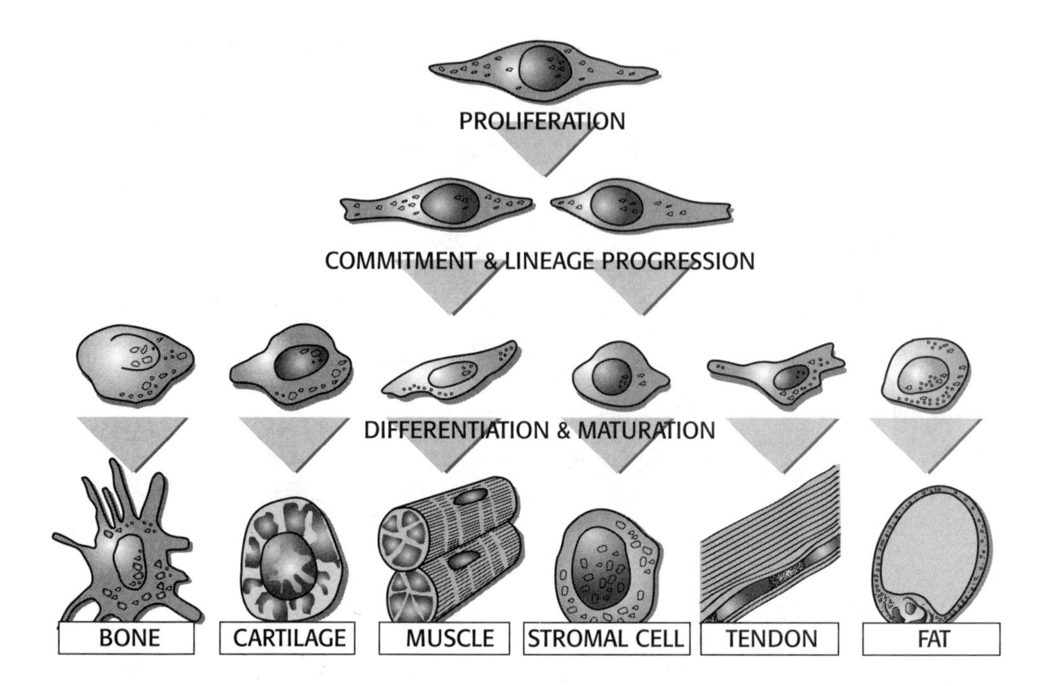

FIGURE 6.8

The promise of regenerative medicine.
Adapted, courtesy of BioWhittaker, Inc.

beginning to see their formerly disparate fields as one system," says Goodell. One compelling general question is whether the pockets of stem cells in adult organs developed as part of the organ or were seeded, as a sort of generic stem cell army, during embryogenesis. Marking cells and following them as an animal develops should answer that question.

Right now, researchers are reconfirming what at first seemed impossible—that blood can become brain, and brain can become blood; that blood can become liver, and muscle can become blood. Then they will discover the conditions necessary for these events to occur and will reveal other interconversions. The fact that these first interconversions all implicate bone marrow may simply reflect the ease of working with this highly accessible tissue.

The next step—and it will be a huge one—will go beyond cell biology to decipher how cells and tissues form organs, then attempt to recreate this dance of development either outside or inside bodies. As developmental biologists learn the particular combinations of biochemicals, cell densities, mechanical forces, and spatial organizations necessary to nudge development in specific directions, applications will follow closely behind. It's exciting to imagine the regenerative medicine of the future, something that will surely be in demand as the population ages. Perhaps one day the human body will be ultimately mapped out to indicate the pockets of developmental potential that persist into adulthood, revealing reservoirs that medical scientists can tap to enable us to heal ourselves.

REFERENCES

Bartlett, Perry. "Pluripotential hematopoietic stem cells in adult mouse brain." *Proceedings of the National Academy of Sciences* 79:2722–2725 (1982).

Bjornson, Christopher R. R., et al. "Turning brain into blood: A hematopoietic fate adopted by adult neural stem cells *in vivo*." *Science* 283:534–537 (1999).

Bradley, Allan, et al. "Formation of germ-line chimaeras from embryo-derived teratocarcinoma cell lines." *Nature* 309:255–257 (1984).

Clarke, Diana L., et al. "Generalized potential of adult neural stem cells." *Science* 288:1660–1663 (2000).

Doetschman, Thomas, et al. "Targetted correction of a mutant HPRT gene in mouse embryonic stem cells." *Nature* 330:576–577 (1987).

Doetschman, Thomas, et al. "Establishment of hamster blastocyst-derived embryonic stem (ES) cells." *Developmental Biology* 127:224–227 (1988).

Gussoni, Emanuela, et al. "Dystrophin expression in the *mdx* mouse restored by stem cell transplantation." *Nature* 401:390–394 (1999).

Jackson, Kathyjo Ann, et al. "Hematopoietic potential of stem cells isolated from murine skeletal muscle." *Proceedings of the National Academy of Sciences* 96:14482–14486 (1999).

Lewis, Ricki. "A paradigm shift in stem cell research?" *The Scientist* 14:1 (2000).

Martin, Gail R. "Isolation of a pluripotent cell line from early mouse embryos cultured in medium conditioned by teratocarcinoma stem cells." *Proceedings of the National Academy of Sciences* 78:7634–7638 (1981).

Matsui, Yasuhisa, et al. "Effect of steel factor and leukaemia inhibitory factor on murine primordial germ cells in culture." *Nature* 359:550–551 (1992).

McKay, Ronald. "Stem cells in the central nervous sytem." *Science* 276:66–71 (1997).

Petersen, B. E., et al. "Bone marrow as a potential source of hepatic oval cells." *Science* 283:1168–1172 (1999).

Scheffler, Bjorn, et al. "Marrow-mindedness: A perspective on neuropoiesis." *Trends in Neurosciences* 22:348–357 (1999).

Shamblott, Michael J., et al. "Derivation of pluripotent stem cells from cultured human primordial germ cells." *Proceedings of the National Academy of Sciences* 95:13726–13731 (1998).

Solter, Davor, and John Gearhart. "Putting stem cells to work." *Science* 283:1468–1472 (1999).

Stevens, Leroy C. "Studies on transplantable testicular teratomas of strain 129 mice." *Journal of the National Cancer Institute* 20:1257–1270 (1958).

Stevens, Leroy C. "Embryology of testicular teratomas in strain 129 mice." *Journal of the National Cancer Institute* 23:1249–1294 (1959).

Stevens, Leroy C. "The development of transplantable teratocarcinomas from intratesticular grafts of pre- and postimplantation mouse embryos." *Developmental Biology* 21:364–382 (1970).

Theise, Neil D., et al. "Derivation of hepatocytes from bone marrow cells in mice after radiation-induced myeloablation." *Hepatology* 31:235–240 (2000).

Thomas, Kirk R., and Mario R. Capecchi. "Site-directed mutagenesis by gene targeting in mouse embryo-derived stem cells." *Cell* 51:503–512 (1987).

Thomson, James A., et al. "Isolation of a primate embryonic stem cell line." *Proceedings of the National Academy of Sciences* 92:7844–7848 (1995).

Thomson, James A., et al. "Embryonic stem cell lines derived from human blastocysts." *Science* 282:1145–1147 (1998).

Watt, Fiona M., and Brigid L. M. Hogan. "Out of Eden: stem cells and their niches." *Science* 287:1427–1430 (2000).

THE ROOTS OF CLONING

History is built around events that are said to divide time. The cloning of a mammal from the nucleus of an adult's cell is one such event, at least in the public eye. Cloning time is unofficially defined as "before Dolly" and "after Dolly," referring, of course, to the famed Scottish sheep whose every cell descends from the nucleus of a cell from a 6-year-old donor ewe.

Cloning has captured attention, spurred imagination, and catalyzed fears as little else has in life science in recent time. Ask a person on the street what a prion, telomere, or archaeon is, and a blank stare is likely to be the response. Mentioning cloning, in contrast, tends to conjure up futuristic images of mass-produced Homo sapiens, *restored dead children, and spare body parts.*

The media have had a limited love affair with cloning in the past, but it never really created more than a ripple of an impact until Dolly. The 1970s gave us Woody Allen's classic film Sleeper, *in which hero Miles Monroe attempted to clone a deceased leader from a cell plucked surreptitiously from his nose. The 1980s took a serious turn with* The Boys from Brazil*'s cloned Nazis, and a 1984* National Examiner *headline proclaimed the inevitable and long-awaited cloning of Elvis. The 1990s gave us Michael Crichton's* Jurassic Park, *where cloned dinosaurs miraculously grew to gargantuan size in just a few years, and the multiple "Eves" of* The X-Files, *the murderously demented multiple offspring of an out-of-control egomaniacal embryologist.*

Dolly wasn't a great surprise to scientists familiar with the elegant experiments that molded developmental biology over the preceding decades. Her creation did, however, transcend earlier work because it began with a nucleus from an adult's cell, albeit an adult three years deceased, her cells frozen and stored away. Neither was Dolly's debut a "discovery" in the sense considered in previous chapters. Cloning certainly does not fall into the "artistic" style of scientific discovery, in which an observer makes unusual connections or goes against all that he or she knows to be true, such as Stanley Prusiner's relentless pursuit of prions. Neither is cloning purely the systematic type of science, in which an investigator doggedly searches for something that evidence indicates must be there—such as sequencing a genome, describing telomeres, or watching stem cells bloom into blood or bone, muscle or cartilage.

Cloning a sheep from an adult cell nucleus was quite the opposite of Woese's discovery of a third form of life or the finding of living communities in deep-sea hydrothermal vents. It was planned. Anticipated. Predicted, if not even certain, to eventually happen. And so Dolly's creation was not so much an illustration of scientific discovery as it was an exercise to see if something could be done. She is the culmination of many years of work that addressed certain fundamental questions. How does a cell access some of its genetic material while silencing the rest of it in order to specialize? Once a cell does specialize, can its nucleus return to its embryonic state, and perhaps be reset? "As an embryo develops, each cell responds to molecular signals that make it more restricted in the range of genes expressed. A cell type has a signature pattern of expressed genes. Cloning requires that the pattern of gene expression be reprogrammed. We are asking the cell to forget the signals and gene expression, and recommence development," says Virginia Papaioannou, a professor of genetics and development at Columbia University.

Early experiments on amphibians began to answer these questions. In a sense, the cloning work of the 1990s is not raw, new discovery, but confirmation of the hypothesis that one can, indeed, turn back the developmental clock.

THE AMPHIBIAN YEARS

Clones are common. A bacterium that divides unchecked spawns a population of identical millions within days. A grove of poplars may be clones linked underground, as are fungi. Aphids typically spawn several generations of armies of identical females before entering a sexual stage. An armadillo produces four clones each time it gives birth.

Clones aren't magic. They are merely nature's version of "If it ain't broke, don't fix it." When the environment is favorable—abundant food, no predators, mild weather, no natural disasters—it's okay if every individual in a population of organisms is exactly the same genetically. This is the essence of asexual reproduction—mass-producing a genome that suits the prevailing conditions. Of course asexual reproduction is a trade-off of present comfort against possible future doom. Should the environment change in a way that harms the complacent genome, sameness becomes a liability and all die. In contrast, sexual reproduction fuels population diversity. Faced with a plague, individuals blessed with better immunity survive, persist, and repopulate. In the face of a new predator, those with longer legs or more mobile seeds might survive. At the population level, diversity ensures safety.

The researchers who first attempted cloning were not interested in creating great plagues of identical frogs. They were interested in learning about just the opposite—how cells come to be different from each other. With their large, numerous, and easily accessible eggs, amphibians are ideal model organisms. One can easily observe the choreography of early embryogenesis in an amphibian by scooping the tiny balls of cells from water and observing them under a microscope.

The first person to suggest the "somewhat fantastical experiment" of cloning an organism from a somatic cell nucleus was German zoologist Hans Spemann. He not only originated the idea, but also actually carried out the first bona fide cloning experiment—doing it in 1928, then formally describing it a decade later in his book *Embryonic Development and Induction.* But like all scientific leaps, Spemann's hardly sprang from just his imagination. The creative spark came from others, most notably August Weismann, a zoologist and anatomist from the University of Freiberg.

Weismann worked and thought and taught in 1892, that slice of time after Darwin yet before the turn-of-the-century rediscovery of Gregor Mendel's paper that derived and explained the laws of inheritance. In his book *The Germ Plasm: A Theory of Heredity,* Weismann proposed precisely what cloning has disproven—that as development proceeds, cells specialize by gradually losing some of their genes (although at the time the term *gene* was not yet in use). Development was seen as a continual restricting of potential, with possibilities permanently turned off as a cell acquired its distinctive characteristics. He wrote, "In each nuclear division the specific plasm of the nucleus would be divided, according to its nature, into nonequal moieties, so that the cell bodies, the character of which is determined by the nucleus, would thus be newly stamped." Spemann read Weismann's work while in a sanitarium recovering from tuberculosis during the winter of 1896–97. It was a little like Charles Darwin devouring Sir Charles Lyell's *Principles of Geology* while stricken with seasickness aboard the H.M.S. *Beagle* in the Galapagos Islands.

A spate of experiments followed in the wake of Weismann's book. Various researchers using various organisms teased apart the partners of two-celled embryos and watched twins develop. This demonstrated the totipotency of cells of a two-celled embryo (that is, the ability of each cell to support the development of a whole organism), but it did not reveal the developmental restriction that Weismann proposed. One such experiment would inspire Spemann. Jacques Loeb, at the University of Chicago, noted that sea urchin eggs—easy to work with because they are parthenogenetic—occasionally form a dumbbell-like shape, with a small ballooning of cytoplasm off to the side. Sometimes, when the cell divided, one of the daughter nuclei would scoot into the bleb of cytoplasm, which would then pinch itself off, and twins would develop.

Thinking of the sea urchin eggs, Spemann developed a novel technique with the help of a hair from his newborn son. He used the hair to create a tourniquet to tie off the parts of a two-celled salamander embryo. Not surprisingly, twins resulted. He then became slightly obsessed with a variation on the theme in which he did not completely sever the cells, which resulted in the development of "strange double creatures"—salamander Siamese twins. He repeated this feat some 1500 times. Then he tried a version of Loeb's experiment. He tied the baby hair around a fertilized egg in a way that pinched off a bleb of cytoplasm, letting the other side, which contained the nucleus, divide four times, yielding a ball of 16 cells. When he loosened the tourniquet slightly, one of the nuclei migrated into the pinched-off cytoplasm. He pulled the microscopic string, and the bleb and its new nucleus went on to develop into a salamander.

It was, in essence, the first cloning experiment, by a technique that would become known as embryo splitting—just done recently on rhesus monkeys (Figure 7.1). But Spemann realized that a true test of Weismann's hypothesis would be to somehow obtain donor nuclei from cells of an adult, or at least from more mature and specialized tissue. He could envision the "what," but not the "how," of the procedure, writing in his 1938 book, "The first half of this experiment, to provide an isolated nucleus, might be attempted by grinding the cells between two slides, whereas for the second, the introduction of an isolated nucleus into the protoplasm of an egg devoid of a nucleus, I see no way for the moment."

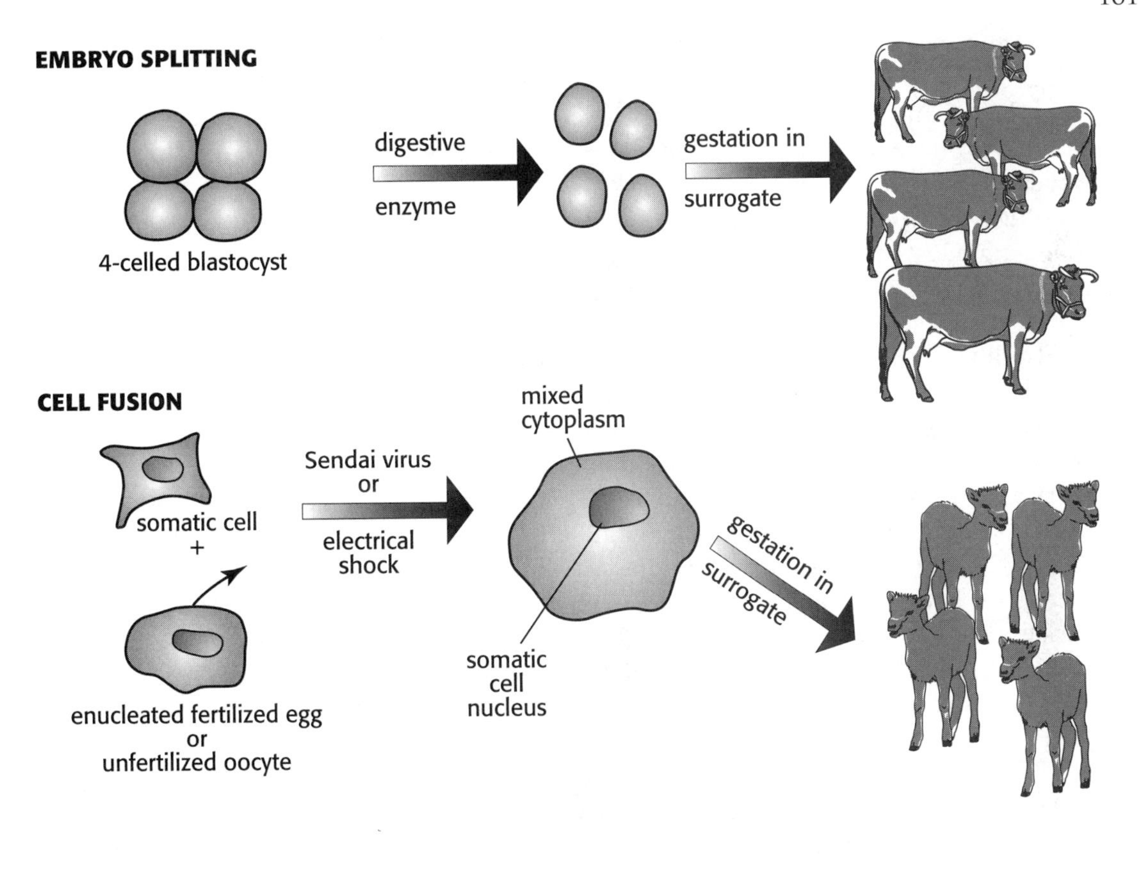

FIGURE 7.1

Three routes to cloning.

A decade later, technology caught up with Spemann's ideas. Embryology had evolved from a largely descriptive and even metaphysical discipline to a more rigorous experimental science. Researchers began to manipulate embryos rather than passively watching development unfold and wondering at the underlying mechanisms. And so the saga of testing whether a nucleus from a somatic cell could support development picked up again, at the Institute for Cancer Research in Philadelphia, in the laboratory of Robert Briggs.

The experimental subject in the Philadelphia lab was *Rana pipiens,* the leopard frog. Besides the ease of obtaining these frogs and their jellied eggs from nearly any pond, the eggs had the useful ability to start developing following a brief pinprick. Leopard frogs normally progress through three tadpole stages, each more specialized than the last: tail-bud, heartbeat, and swimming.

Briggs knew what he wanted to do, but didn't know how to do it. So he found an assistant, a graduate student at New York University, Thomas King, who had mastered the technique of creating needles and micropipettes from glass, then plucking out nuclei and transferring them to other cells. After a few grant rejections from the National Institutes of Health, which were based on reviewers declaring the idea of cloning absurd and impossible, a site visit by the agency finally secured funding to bring aboard the young King as a research fellow in early 1950. The work began.

The first series of experiments took donor nuclei from embryos consisting of 8000 to 16,000 cells. These were hollow balls of cells, called *blastocysts,* that had not yet set aside the sheets of cells that would ultimately fold into the tissue layers and organs of a bona fide embryo. King transferred nuclei from the cells of the blastocysts—called *blastomeres*—into fertilized eggs shed of their own nuclei. At first these new cells would support only a few cell divisions, but finally, in November 1951, one lone embryo kept going—until someone accidentally smashed it with a forceps. King repeated the feat a month later, and in 1952, the first famous Briggs and King paper appeared, recounting a 60% success rate in coaxing blastomere nuclei to support development to the feeding tadpole stage.

By 1956, data had amassed. Of 197 vacated fertilized ova given new blastomere nuclei, 104 divided, 35 became embryos, and 27 survived until tadpolehood. Although these were certainly good odds, especially compared with Dolly's 1 in 277 statistic, Briggs and King knew that their success was probably due to the early stage of the nucleus donors. Similar experiments using nuclei from a later-stage embryo, the fully folded gastrula, had been performed in 1955 with no success at all. And so Weismann could still be correct: Fully differentiated cells could have some of their genes jettisoned or at least silenced, perhaps irreversibly.

In the next few years, others replicated Briggs and King's findings, in other species. But they all found the same limitation: The older the donor nuclei, the harder it was to clone. Two explanations emerged. Either Weismann was correct and nuclei from specialized cells were permanently restricted in developmental potential, or the act of removing nuclei and transferring them to new surroundings so stressed cells that they could not continue development very far. Later, researchers would learn that nuclear transplantation often breaks chromosomes, perhaps a consequence of asking a nucleus no longer accustomed to rapid division to race through the process in its newly embryonic surroundings, a little like asking a 50-year-old runner to compete against 18-year-olds. The key experiment would be to keep trying, with perhaps a different technique, to see if a nucleus from an adult's cell could tap its potential or if that potential was truly gone.

An Oxford University developmental biologist, John Gurdon, picked up the train of experimentation. He used *Xenopus laevis,* the African clawed frog, which was a laboratory favorite because it raced through development three times faster than *Rana pipiens. Xenopus* is a more primitive frog, lacking the tongue, ears, and eyelids of *Rana.* Plus, it can readily regenerate its limbs.

In 1962, Gurdon reported that he had coaxed nuclei from intestinal lining cells of feeding tadpoles to support development of adult frogs. It was difficult to do, with about a 2%

success rate, but one cloned frog was enough to demonstrate that a nucleus from a differentiated cell could indeed be reactivated. By 1966, he'd gotten cells from tailbud larvae to support development to adulthood too. But as expected when someone succeeds at a goal that has eluded others, doubters stepped forward. A student of Briggs', Dennis Smith, suggested that perhaps Gurdon had inadvertently picked up not intestinal lining cells, but the primordial germ cells (sperm or eggs to be) that migrate through the gut during larval development. The primitive nature of these cells would explain why they were able to support development. Smith's challenge foreshadowed the doubt that would greet the announcement that a sheep had been cloned 30 years later. The young man based his claim on experiments he had performed on different species whose primordial germ cells did not wander through the tadpole's gut. Using such animals, he was unable to replicate Gurdon's results.

But in science as in other aspects of life, when something works and no one believes it, the way to show that it is valid is to do it again. In 1975, Gurdon reported another series of experiments that supported his earlier findings that limited cloning was possible in amphibians, but difficult to achieve. This time, he transplanted nuclei from keratinized (hardened) skin cells taken from the webbing between a frog's foot into enucleated eggs. Development proceeded, but only as far as the last tadpole stage.

Gurdon's work was widely hailed as being an important step toward cloning. But it was, in a sense, incomplete. He had accomplished the feat from a larval (tadpole) donor to support adult development, and from an adult donor to support larval development. He had not shown that a nucleus from an adult's cell could support the entire developmental journey, from adult cell nucleus to adult cloned animal. That would be Dolly's contribution.

THE MOUSE, TAKE 1

The trail of cloning was soon to segue from amphibians to mammals. Transplanting nuclei from amphibians was easy by comparison. The eggs of mammals are tiny and rare, not floating in pond scum. Moreover, mammalian cytoplasm is not as accepting of foreign nuclei, the time of the new organism's genome activation varies from species to species, and mammals must have a genome from a male and a female to develop normally. If these hurdles weren't enough, mammalian embryos tend to nestle into hard-to-reach uterine linings—they do not dance through their stages in water for all to see, as the comparatively cooperative amphibian embryos do. Still, cloning briefly attracted media attention with the 1978 publication of a book by David Rorvik, called *In His Image: The Cloning of a Man,* which claimed that the feat had been accomplished in South America. The account was fictional, but it got the public thinking seriously about cloning.

Back in the scientific camp, in the years following Gurdon's demonstration of cloning in frogs, many developmental biologists attempted to repeat the experiments using mammals. A brief flurry of excitement surrounded work published by renowned University of Geneva biologist Karl Illmensee, who, with Peter Hoppe of the Jackson Laboratory in Bar Harbor, Maine, claimed in early 1981 to have cloned three mice from the nuclei of early embryo cells. But when others could not replicate the experiments, and people began to realize that no one had ever actually seen Illmensee demonstrate his technique, doubt was cast.

Meanwhile, Davor Solter, then a developmental biologist at the Wistar Institute in Philadelphia, with postdoctoral research associate James McGrath, was trying to clone mice using an easier technique than injecting tiny mammalian eggs. He used a virus called Sendai that causes cells to fuse their cell membranes, becoming one. First, he fused a fertilized mouse egg whose nucleus had been removed with another fertilized egg, and implanted it into a surrogate mouse mother; a pup developed. This experiment showed that the technique worked. But when Solter fused an enucleated fertilized mouse egg with a cell from a two-celled embryo, the organism wouldn't develop beyond a hollow ball of cells, a blastocyst. He tried it over and over, and it just wouldn't work.

Solter faced a dilemma, because in science, no news is bad news. Journals do not like to publish negative evidence. Granting agencies do not like to see evidence that something did not work. But Solter published the findings first in *Cell* in May 1984, then in *Science* in December. The *Science* paper ended with a sentence that would place a pall over cloning research for years: "The cloning of mammals, by simple nuclear transfer, is biologically impossible."

But of course cloning a mouse turned out not to be impossible, and Solter would have the honor of writing an article that would accompany announcement of the accomplishment. It would be done in 1998, after researchers had perfected nuclear transfer into oocytes (immature unfertilized eggs) and learned to coordinate the cell cycles of donor nucleus and recipient cytoplasm. Recalls Solter, today the director of the Max Planck Institute for Immunobiology in Freiburg, Germany, with the benefit of hindsight, "We used a fertilized egg, a zygote, as a recipient. At that time it seemed logical to start with an egg that has been correctly activated. But now we know that it is better to start with the oocyte and not a fertilized egg."

OF BOVINES AND OVINES—COWS AND SHEEP

McGrath and Solter's public declaration of the impossibility of cloning put a damper on research in academic circles, but interest persisted in another arena—agriculture; in particular, investigators working to improve traits in large domesticated animals such as cows, sheep, goats, and pigs. Whereas the amphibians and mice that had been the subject of cloning attempts to date were chosen for their value as model organisms and were used to unveil basic biological principles, agricultural research was, of course, more directed, with more immediate commercial goals.

On the bovine front were Neal First and his colleagues at the University of Wisconsin, including postdoctoral researcher James Robl and doctoral student Randall Prather. Funding came from W. R. Grace and Company, which was interested in propagating high-value animals. Robl left for the University of Massachusetts, Amherst, and young Prather eventually cloned a calf from a nucleus of an early embryo cell after mastering a technique developed by Danish researcher Steen Willadsen.

The approach was to isolate a sheep's oviduct and grow it in the laboratory, where it provided a living bioreactor of sorts to nurture temporary embryonic residents. (Sheep oviducts were easier to obtain and maintain than cow oviducts, and the embryos within seemed not to care about the origin of their surroundings.) Very early cow embryos did well in the isolated sheep oviducts. Prather used digestive enzymes to separate the components of eight-celled cow embryos, then applied an electrical pulse, rather than Sendai virus, to fuse embryo cells with enucleated oocytes. Following incubation in the disembodied sheep oviduct, the cloned embryos completed gestation in a surrogate cow. Prather and First published their paper in the November 1987 issue of *Biology and Reproduction.* A cloned cow had officially entered the scientific literature.

Meanwhile, on the ovine front, Willadsen was working at the British Agriculture Research Council's Unit on Reproductive Physiology and Biochemistry in Cambridge. He preferred sheep to cows. The animals were smaller, had a shorter gestational period, and were more plentiful and therefore cheaper to experiment with than cows. In addition to developing the sheep oviduct method of nurturing early embryos, Willadsen in 1979 described in *Nature* his technique of placing agar around isolated oocytes to compensate for the loss of the surrounding protective layer, the zona pellucida. Unfertilized eggs shorn of their zonas would often die from being manipulated. The agar coat not only protected the delicate cells, but also adhered and stayed in place, providing an easy way for Willadsen to identify cells later on.

Willadsen succeeded first in separating a series of two-celled embryos into single cells and seeing how far they would develop—similar to the experiments done in Weismann's time. Of 35 cleaved embryos, 61 of the 70 cells developed as far as blastocysts, at which point he ended the experiment. The next set of experiments extended the timetable, resulting, ultimately, in the births of five sets of twins and five single lambs from separated two-celled embryos. But when Willadsen started with older embryos, he quickly discovered that the success rate plummeted with the increasing age of the donor embryos. Whereas from 60% to 80% of separated two-celled embryos survived, the number fell to 50% for separated four-celled embryos, and to 5% to 10% for separated eight-celled embryos. Along the way Willadsen combined blastomeres from a sheep and a goat and raised it in a surrogate ewe, resulting in a bizarre hybrid sporting horns and hair from both species that graced the cover of *Nature.*

To increase the success rate, Willadsen switched from embryo splitting to using Sendai virus or an electric shock to fuse somatic cells with enucleated oocytes. He developed a way to split oocytes, and used the halves lacking nuclei in the cloning experiments. On March 28, 1984, he deposited in one such vacated oocyte a nucleus from an 8-celled embryo that had been frozen on November 11, 1979. That experiment led to a lamb that appeared in *Nature* on March 6, 1986. At about the same time, First's lab was having similar success on cows but took longer to publish the results, partly because bovine gestation is 9 months versus 5 months for sheep. Willadsen was able to repeat the work on 16-celled embryos, and First's group with 32-celled embryos.

The ability to clone cows and sheep from early embryo cell nuclei resurrected the feasibility of cloning that had been blindsided by McGrath and Solter's failure with mice. Wrote Willadsen in the 1986 *Nature* paper, "[A] firm basis has been established for further experiments involving nuclear transplantation in large domestic species."

Industry recognized the dual milestone, and so began an effort to commercialize transfer of early-embryo nuclei to clone valuable livestock. Willadsen was wooed to Granada Genetics in Houston to clone cows. For awhile, it seemed as if the companies themselves were undergoing a cloning process, with Genmark forming in Salt Lake City, Alta Genetics in Alberta, and W. R. Grace supporting Infigen in DeForest, Wisconsin.

In its heyday, Granada scientists scooped out 1000 oocytes a day from ovaries collected from slaughterhouses. Explains George Seidel, a professor of physiology at Colorado State University in Fort Collins, "To clone calves, they would take a 16- or 32-celled embryo and fertilize an enucleated egg with one nucleus from it. The eggs would get the diploid gene complement from one cell of an embryo, rather than half from a sperm and half from an egg." Granada used Willadsen's oocyte-halving and sheep oviduct technique. "The good ones were placed in cows, one or two at a time. Of 32 embryos starting out, maybe 3 or so would be okay to put into cows. This is done with cell culture of oviduct linings now. But in the days of Granada, they had to use nature, a sheep. An oviduct is a good place to culture embryos," Seidel adds.

Ultimately, calf cloning proved too expensive to attract a market and, combined with unrelated business developments, led to Granada's demise in 1991, says Mark Westhusin, a nuclear transfer specialist at Texas A&M University in College Station who heads a program to clone a dog. He was a research scientist at the Granada Biosciences Inc. arm of the company from 1986 until 1991. "Granada closed because of a complicated series of issues, financial problems. They were trying to go public, and tried to reorganize. Granada was huge. It included cattle ranches, a shrimp farm in Panama, and Texas meat purveyors. Cloning was just part of it."

Seidel recalls one unanticipated problem that affected Granada and persists today: "large calf syndrome." Clones are oversized, hampering their ability to survive the neonatal period. They grow to huge proportions because the placentas are abnormal. This may be a consequence of starting life with one parent, for it is a male genome that normally oversees development of extraembryonic structures. A clone derives from a somatic cell, not a male and female genome joining. The fact that the male genome controls placental development illustrates a little-understood phenomenon called genomic imprinting. During development, certain clusters of genes from one parent are shut off by shields of methyl groups, which enables the contribution from the other parent to predominate. Apparently, normal development depends on imprinting of certain genes from certain parents, as if the formation of a sexually reproducing organism is a carefully choreographed dance with each partner carrying out specific steps. With each generation, as sperm and ovum form, the imprints are erased, so that they can be reset. Cloning skips this imprinting stage, and one result may be an overgrown placenta—and an oversized animal.

No one will directly implicate failed cloning in Granada's end, but the resulting cattle were reportedly not normal. When the company disbanded, Seidel's department purchased 73 of the pregnant cows to use their cloned cargo in nutrition experiments. They soon found that most of the calves did not survive for very long—more than a third died unless they were cared for very intensely as newborns. Seidel feels that still-unidentified metabolic problems caused the low survival rate, rather than the large size. The culprit could have been any of many nonnatural manipulations—embryo freezing, life in an oviduct of another species, the one-parent problem, or prolonged time in culture. Somatic cells can only divide from 40 to 60 times in culture, a restriction called the Hayflick limit (see Chapter 5). Toward the end of that lifetime, errors in DNA replication may occur that can compromise development, because such a cell would never normally become the first cell of a new individual.

The practice of serial cloning also may have been simply too much to maintain normal development. "They would take a 32-celled embryo and get 32 identical cells, and use the nucleus of each to fertilize an egg. They'd let those get to the 32-celled stage—about six of them would make it—and then repeat the process. That would give 32 multiplied by 6—192—identical cells whose nuclei would be used to fertilize 192 eggs. The problems with culturing embryos are exacerbated as you go to serial cloning," Seidel explains.

Still, the large size has remained a concern. "I'll bet there were plenty of veterinarians around at Dolly's birth to cope with any problems," he adds.

By the early 1990s, the calf cloning industry was pretty much finished. What little interest in cloning that remained in academic quarters following the 1984 McGrath/Solter bashing faded even further as two other technologies captured attention and grant funding. Transgenic technology introduced selected genes into fertilized ova or sperm or eggs, enabling researchers to create animals and plants with desired specific traits without worrying about duplicating entire genomes. Cloning began to seem unnecessary. The death knell may have been the advent of so-called knockout technology, which is a targeted version of transgenesis. Stem cells with engineered changes are mixed with early embryos, and then the resulting animals are bred to create individuals with the desired alteration stitched precisely into each cell (see Chapter 6).

Yet even as the cloning companies were bowing out, the prospect of human cloning reared its head briefly, a peek ahead, perhaps, to what would happen when Dolly's existence was announced a few years later. At a scientific meeting in 1993, George Washington University researchers Jerry Hall and Robert Stillman reported that they had performed "blastomere separation"—also known as embryo splitting—on several human two-celled embryos. Their source material consisted of "rejects" from fertility clinics, not slated for use because each had been fertilized by more than one sperm, a developmental disaster. So the researchers gently separated some of the cell pairs and watched the cells continue through a few cell divisions (Figure 7.2).

The culmination of Hall and Stillman's experiments was balls of cells, albeit human balls of cells. The media saw it differently. Soon, headlines trumpeted human cloning, with the standard *Jurassic Park* and *Brave New World* comparisons. Hall, Stillman, and scores of scientists attempted to point out that this was not, technically, cloning, but rather a version of assisted twinning, to little avail. The story grew old.

MEGAN AND MORAG PROVE THE PRINCIPLE OF CLONING—BUT ARE IGNORED

While it seemed that the rest of the world had been sidetracked from cloning, Ian Wilmut, a researcher at the then-little-known Roslin Institute located about 7 miles from Edinburgh, in Midlothian, remained intrigued. He had worked on freezing boar sperm—an important endeavor because boars release semen a pint at a time—and on freezing embryos, to study the causes of spontaneous abortions. Back in 1986, Wilmut had learned of Willadsen's unpublished work on cloning from 60- and 120-celled embryos, and he set out to push the boundaries of developmental potential even further. But by the early 1990s, Wilmut became enamored with transgenics and knockouts to introduce and propagate precisely a gene of interest.

The duo of technologies were powerful tools for studying genetics, and at the same time offered a way to tailor domesticated animals and plants to produce desired proteins. By 1991, Wilmut had achieved brief notoriety by creating a flock of transgenic sheep that secreted α_1-antitrypsin in their milk, a human protein useful in treating hereditary emphysema. Both transgenic technology and knockouts were technically difficult, however. Could cloning, wed to either technique of introducing genes of interest, be used to mass produce genetically engineered organisms? It might eliminate steps and increase efficiency. "With nuclear transfer, you can make all animals in one generation (sheep and cows),

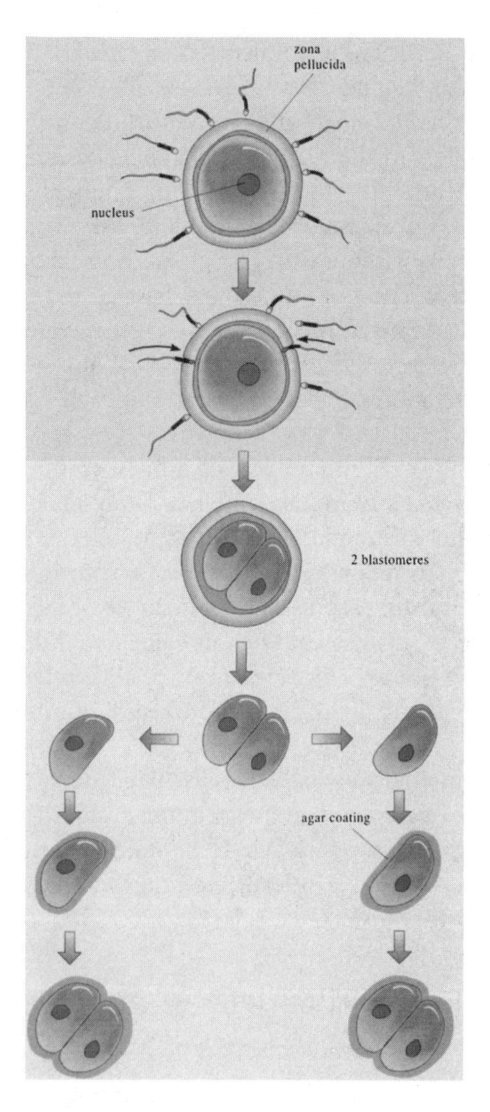

FIGURE 7.2

Embryo splitting is not cloning by nuclear transfer. When two George Washington University biologists did this in 1993, they sparked headlines that they had cloned humans. In early 2000, a team from the Oregon Regional Primate Research Center similarly reported embryo splitting of 107 rhesus monkey embryos, yielding 368 individual animals. The first monkey born was named Tetra because she came from a split four-celled embryo. This technique mimics natural formation of identical multiples.

Lewis R. Human genetics: concepts and applications. *3rd ed. Boston: McGraw-Hill, 1999:57. Reproduced with permission of The McGraw-Hill Companies.*

and it would take two and a half years to get milk," according to Alan Colman, research director at PPL Therapeutics Inc., a firm in Midlothian associated with the Roslin Institute. "'Transgenic pharming' without cloning takes four and a half to five years. Plus, cloning could significantly improve upon the typical 0.5 to 3 percent rate of integration of the desired gene into the recipient cell's genome via transgenesis," he added.

Even if cloning—by nuclear transfer or cell fusion—were to become routine or even automated in agriculture, the laws of nature would still intervene, because genetic uniformity is not a typical state of affairs in a natural animal population. Cloning is asexual. Propagating a valued animal would require not herds of he-clones and she-clones who would mate and perpetuate themselves, but instead a setup in which nuclei could be easily harvested from somatic cells and placed in enucleated unfertilized eggs. If mating were to become part of the scheme—if male and female clones were to be crossed—then each of the animals' thousands of genes would have to be homozygous. That is, the two copies of each gene would have to be identical. If not, then meiosis would intervene, mixing up traits in the next generation. If any gene was heterozygous, the animal's sperm or eggs would not be identical because of Mendel's law of segregation—alleles separate during meiosis and are partitioned into different gametes. The cloning would go right out the window.

Still, cloning can replace or be combined with various other ways to reproduce valued animals. Says Steve Stice, associate professor in animal and dairy science at the University of Georgia and cofounder with James Robl of Advanced Cell Technology Inc. in Shrewsbury, Massachusetts, a firm that clones cattle, "Cloning allows us to introduce genetic change faster than through artificial insemination. We can predict an increase in productivity by selecting superior animals and mating them to other clones. Cloning will have a place in agriculture, but not all animals will be clones, because we have to use traditional breeding to introduce new variants." Artificial insemination is slower than cloning because the contribution of the female must be either desired or bred out.

Despite the inherent limitations of cloning, Wilmut, did, of course, go on to become Dolly's "father." But that might not have happened without the help of a young doctoral student in veterinary science in the Roslin lab, Lawrence Smith. It was Smith who realized that the poor efficiency of cloning to date might stem from asynchrony between the donor nucleus and the recipient cytoplasm—that is, the two components of a clone-to-be were at different stages of the cell cycle, the sequence of events that a cell experiences as it prepares to divide (Figure 7.3). Explains Davor Solter, "You have to visualize what exactly happens when they put the nucleus in. The nucleus that gets into the oocyte has a diploid set of chromosomes, but the DNA has not been replicated. Normally what would be in the oocyte at that time is a haploid nucleus. It still has to undergo the second meiotic division. The chromatids have to separate. But the nucleus put in doesn't have chromosomes that can separate. Normally the cell would try to complete the second meiotic division, and throw away a certain number of chromosomes as the polar body.

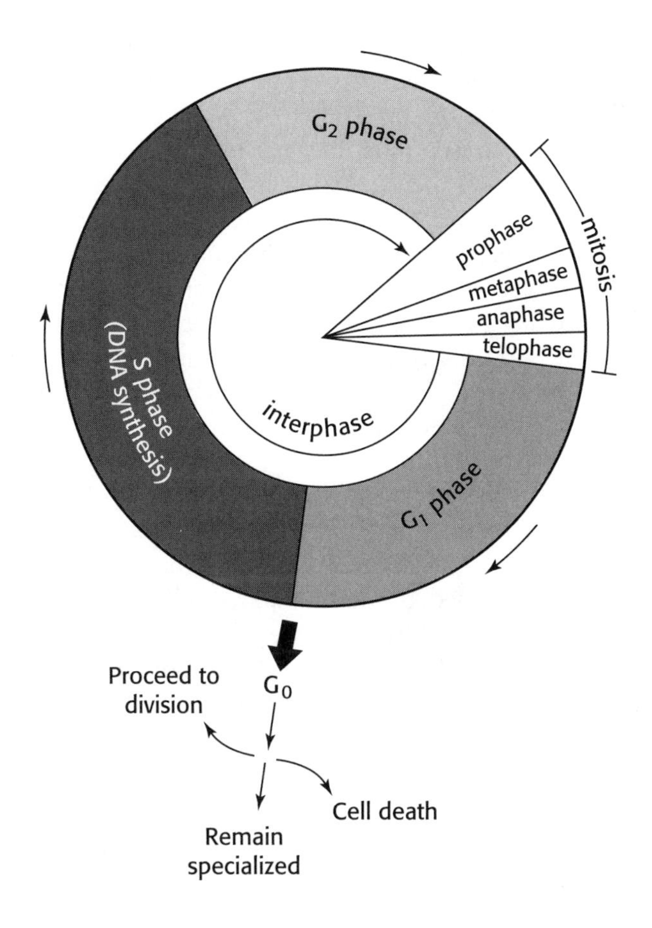

FIGURE 7.3

The cell cycle. At first cloning success seemed dependent on culturing donor nuclei in the G_0 phase. Several variations on the theme are now possible.

This would preclude further successful development. So you have to prevent completion of meiosis 2 and keep the chromosomes in, and nudge them gently into a round of DNA synthesis. The whole natural order is broken."

Smith left to pursue veterinary interests, and Wilmut hired Keith Campbell to investigate the cell cycle hypothesis. In graduate school Campbell had worked on the cell cycle in frogs, inspired by Gurdon's work on amphibian cloning. He was convinced that Gurdon had shown that many combinations of donor nuclei and egg cytoplasm could be coaxed to develop into a clone. So he kept track of the cell cycle stages of various potential donor nuclei. Starving cells in culture by removing the serum that usually nourishes them could induce them to enter a cell cycle "time out" stage called G_0.

The cell cycle consists of two major stages: an interphase when a cell is preparing to divide, and division itself, the M phase. Interphase consists of two "gap phases" of preparation that bracket the S phase, when DNA is replicated. G_0 is an offshoot of the cell cycle, a quiescent period. Lawrence Smith had hypothesized that somehow this resting time prepared the nucleus to join an egg's cytoplasm. Maybe during G_0 a cell reprograms its DNA before differentiation ensues, a little like a fertilized egg reprogramming the genetic contributions of the two founding cells.

In a preliminary series of experiments, Wilmut and Campbell grew cow embryo cells in culture (sheep cells were temporarily unavailable), starved them into entering G_0, and then transferred the nuclei into enucleated eggs. They were astounded when development continued as far as the blastocyst stage, at which point they halted the experiments. Some of the donor cells had already differentiated into fibroblasts, which are connective tissue cells that have distinctive proteins and extensions. Many experiments showed that the technique of starving donor nuclei into entering G_0 worked, even on more specialized donor cells. It was the third major refinement of the cloning process (Table 7.1).

What many scientists consider to be the definitive demonstration of the feasibility of mammalian cloning actually preceded Dolly by a year. Wilmut and Campbell, with their colleagues J. McWhir and W. A. Ritchie, sent the seminal paper to *Nature* in November 1995. The report ran in the March 7, 1996, issue, and was a birth announcement for Megan and Morag, stars of the article entitled "Sheep Cloned by Nuclear Transfer from a Cultured Cell Line."

Megan and Morag were white, curly-haired Welsh mountain sheep that had been gestated in Blackface surrogate mothers. They and others began as the nuclei of cells from day 9 embryos that had divided 7 to 13 times in culture, where they specialized into fibroblasts. Then the serum was removed, promoting quiescence, and the nuclei transferred to enucleated oocytes. When the clones reached the morula or blastocyst stage, they were implanted into the surrogates. Of five pregnancies that took, five lambs were born. Two died within minutes of birth, however, and a third perished a few days later of a common

TABLE 7.1

Key Steps Forward in Cloning
1. Using agar to replace zona pellucida
2. Using unfertilized oocytes instead of fertilized eggs
3. Synchronizing the cell cycle by starving donor cells into G_0

birth defect, a hole in its heart. Davor Solter contributed the accompanying "News and Views" piece, "Lambing by Nuclear Transfer," happy that he had been disproved by the lambs that he had predicted should never have been gestated, let alone born.

Much of the life science community took note of Megan and Morag, because the two sheep clones arose from nuclei from fetal fibroblasts, which are specialized cells. "The 1997 work that introduced Dolly used the same technique, but rather than using fetal fibroblasts, they used adult cells, and in both cases, the cells came from cell lines. From a scientific standpoint, Dolly is much less interesting than the previous paper, but it was the first clone from adult cells. Despite all the razzle dazzle, it wasn't a particularly big advance over the previous paper," says Seidel. Columbia University's Virginia Papaioannou agrees. "What was important was that they cultured the nuclei prior to cloning. The popular press ignored this, but it caused quite a stir in the biological community."

Indeed, most of the media were oblivious to the significance of Megan and Morag. A few newspapers ran a photo of these first cloned sheep standing nose to nose. And in perhaps the greatest oversimplification of all time, the Associated Press' Malcolm Ritter wrote, "In a feat never before accomplished in mammals, scientists have found a way to turn a laboratory dish full of cells into hundreds of genetically identical sheep." Even *New York Times* science writer Gina Kolata, who would be widely promoted later as "the" reporter to break the Dolly story, admits that she was completely unaware of Megan and Morag and of their importance. The *New York Times* did not mention them at all.

Perhaps the best coverage of these first cloned sheep was an analysis by humorist Tony Kornheiser. He questioned the motivation behind mass-producing beasts that do nothing but continually graze, poop, and jump over fences, suggesting that McDonald's market a cloned sheep dish and that the other three lambs who died suffered a fatal fear of Shari Lewis, the puppeteer who introduced the world to Lamb Chop. But all in all, the 1996 paper didn't raise many, if any, flags.

SHEEP 6LL3: DOLLY

No sooner had Megan and Morag entered the world than Wilmut and Campbell and coworkers were on to the next series of experiments, which would culminate in the birth of sheep 6LL3, better known as Dolly. She was named for country singer Dolly Parton. With the births of Megan and Morag, the commercial arm of the Roslin Institute, PPL Therapeutics, immediately funded Wilmut and Campbell's pursuit of cloning from the nucleus of an adult's cell. So by the time the first *Nature* paper appeared in 1996, Wilmut and Campbell were already passaging the donor cells whose nuclei would become Dolly through the paces of cell culture.

They compared and contrasted the ease of cloning from three sources of nuclei: cells cultured from embryos, fetal fibroblasts, and cells frozen three years earlier from a 6-year-old ewe's udder lining. If anyone had romantic images of a gentle-faced elder ewe fondly looking on as her daughter Dolly gestated in a surrogate, this certainly wasn't the case. "The rumors that Dolly's mom was eaten are wrong," says PPL Therapeutics' Alan Colman. The protocol also varied whether the new early embryos spent their first few divisions in a petri dish on oviduct cells or in the tried-and-true disembodied sheep oviduct of Willadsen, although all the precious embryos derived from the nuclei of adult cells received the cushy oviduct surroundings. Again, the key step was to remove the serum to starve the cells into entering G_0.

TABLE 7.2

Results of the Dolly Experiments

Cell Source	Number Started	Number Retrieved from Oviduct	Number New Embryos	Number Implanted in Surrogate	Number Pregnant	Live Lambs
Poll Dorset embryos	385	231	126	87	15	4
Black Welsh Mountain fibroblasts	172	124	47	40	5	3
Finn Dorset frozen udder epithelium	277	247	29	29	1	1

In normal development, not all fertilized ova make it, and adding the culturing and manipulation of cloning drops the odds of success even further. Table 7.2 shows the data from the three types of experiments. Overall, the massive effort yielded eight live lambs, although one, from the fibroblast group, died soon after birth.

The great event—the birth of Dolly on July 5, 1996—was somewhat anticlimactic (Figure 7.4). The real excitement had occurred months earlier, when ultrasound images revealed that one of the fetuses from the udder cell nuclei was in good shape. In fact, Wilmut was away on vacation at the time of the birth, his wife insistent that he not spend every night in the lab checking on his sheep, as he had for Megan and Morag.

Wilmut and Campbell delayed publication of the results of this new series of experiments while attorneys filed patents. But word got around. The paper was slated for the February 27 issue of *Nature,* with covergirl Dolly posing as Wilmut offered her a fingerful of peanut butter.

HYPED REPORTING, CALLS FOR "NO CLONE ZONES," AND OTHER OVERREACTIONS

The *Nature* family of magazines offers a service to the media: advance news summaries so that journalists can begin researching their articles and arranging interviews before a research report is published. On Wednesday, February 19, such a release arrived on the computer screens of many science journalists. The four-paragraph message from *Nature* was low key, the first line merely stating that "The lamb on this week's cover was raised from a single oocyte (egg cell), whose nucleus had been replaced with that from an adult sheep mammary gland cell. It may be the first mammal to have been raised from a cell derived from adult tissue."

The news release went on to recognize the other sheep that shared Dolly's birthday, to give credit to the work on Megan and Morag, and most important, to state the original goal—to investigate whether the genome of a fully differentiated cell can support development of a new individual. The news release included Wilmut's phone number and e-mail address, and some reporters followed up, especially those familiar with the earlier work.

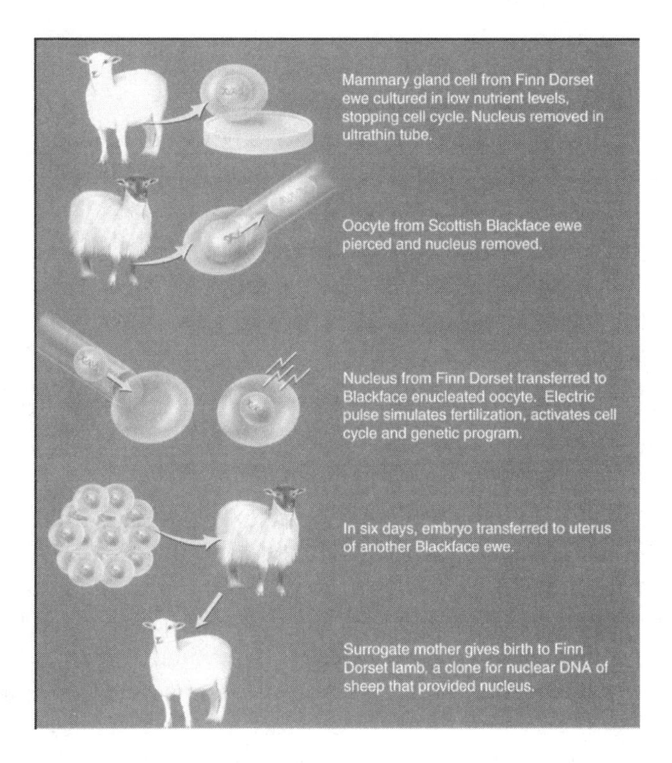

FIGURE 7.4

(a) This is Dolly. (b) Creating Dolly.

(a): Photograph courtesy of PA Photos, Ltd., London, United Kingdom. (b): Lewis R. Human genetics: concepts and applications. *3rd ed. Boston: McGraw-Hill, 1999:32. Reproduced with permission of The McGraw-Hill Companies.*

TABLE 7.3

Practical Applications of Cloning
Speeding the development and uniformity of transgenic herds that carry agriculturally useful traits (meat and milk quality, rate of weight gain, disease resistance, reproductive traits) or that produce pharmaceuticals (clotting factors, growth factors, immunochemicals)
Providing spare parts for transplant
Lowering the number of animals required for certain types of research experiments
Providing consistent specimens for training surgeons
Enabling agriculturalists to assess environmental influences on herds
Preserving or increasing populations of endangered species

But not many writers had *The New York Times* behind them. Gina Kolata's article in Monday's *Times* was the first in the United States on Dolly. The headline (written by editors) was accurate—"Scientist Reports First Cloning Ever of Adult Mammal"—but the subtitles catalyzed the coming hype: "Researchers Astounded," and then "In Procedure on Sheep, Fiction Becomes True and Dreaded Possibilities Are Raised." Of course, biologists were hardly "astounded," and unfortunately, the "dreaded possibilities" set the tone for a slew of other articles. Kolata went on to write a book whose cover bragged that she alone had broken the story of Dolly. She hit the lecture circuit, speaking authoritatively about cloning at various scientific meetings.

Claims to fame aside, the public response to Dolly was immediate, initially negative, and, to many scientists, a great puzzle. Why didn't the public go wild over transgenic or knockout organisms? Or Willadsen's sheep/goat chimera? Why had no one noticed Megan and Morag? As bioethicists jumped into the fray to quell people's fears, reportage shifted to full science fiction mode, evoking images of *The Boys From Brazil*'s cloned Nazis and *Jurassic Park*'s rapidly growing cloned dinosaurs. Often lost in the discussion was why Wilmut and Campbell (and Willadsen and Gurdon and Briggs and King and others) had attempted cloning in the first place—to study a fundamental event of nature, the accessing of the genome. Biologists brave enough to face the media head-on found themselves scrambling to defend the research in practical, more understandable terms (Table 7.3).

Public reaction turned silly. Politicians who didn't know a nucleus from a meatloaf called for "no clone zones." Within days of the story breaking, a website from a company called Clonaid, located in the Bahamas, offered to clone a person for $200,000, or merely bank cells for later cloning for $50,000, with the promise that "cloning will enable mankind to reach eternal life." Other websites offered cloning services, with Elvis and Jesus (from blood on the shroud of Turin) leading the pack. Antiscience sentiment (see Chapter 9) rose, while soap opera writers embraced the topic as an escape from the usual amnesiac identical twin abductee stories. Before you could say "deoxyribonucleic acid," *Guiding Light* had a deranged scientist magically whipping up a clone of the character Reva Shayne. But they goofed. They created the Reva clone by activating an egg she had stored, missing the entire raison d'être of cloning. Still, the Reva clone grew up, way too fast, to face the same fate as *Jurassic Park*'s dinosaurs. She aged decades in a matter of months, just in time for May's television sweeps period. Talk about synchronization strategies!

Meanwhile, scientist skeptics questioned whether Wilmut and company had actually cloned from an adult cell, zeroing in on the tantalizing statement in the *Nature* paper that "The phenotype of the donor cell is unknown . . . . We cannot exclude the possibility that there is a small proportion of relatively undifferentiated stem cells able to support regeneration of the mammary gland during pregnancy." The challenge was reminiscent of Dennis Smith publicly doubting Gurdon's cloned frogs three decades earlier. And so a fleeting backlash began. A headline in the March 2 issue of *Time* magazine asked "Was Dolly a Mistake?" But a DNA fingerprinting analysis would eventually silence that challenge in the July 20 issue of *Nature.* "Alec Jeffreys [of the University of Leicester], the father of DNA fingerprinting, found a one in 100 billion chance that there had been a mistake. We would have needed a field of that many sheep. No doubt the *National Enquirer* would say there are not that many sheep on Earth, and therefore Dolly must have come from an alien," quips Alan Colman. So Dolly was indeed a clone. *Time* apparently changed its tune, and its March 10 issue devoted the cover and much of its coverage to Dolly, elevating Wilmut to celebrity status while at the same time referring, yet again, to *Frankenstein, Brave New World,* and *Jurassic Park.*

Things began to quiet down somewhat until a Friday afternoon in early January 1998, at a press briefing held at the University of Chicago to discuss legal and ethical aspects of human cloning. Here, the public fear that someone with delusions of grandeur, contacts, and technical expertise would try to clone a human sprung to life in the form of a 69-year-old physicist named Richard Seed. To an audience who mostly chose to ignore him, Seed announced that he intended to raise funds to start a clinic to clone humans. He had an interesting background. In the 1970s, Seed cofounded a firm that developed embryo transfer techniques for cattle, and in the early 1980s, he took a brief, failed foray into running a fertility clinic in Chicago.

Joe Palca of National Public Radio couldn't resist Seed's obvious perfect casting for the role of the new Dr. Frankenstein and allowed him substantial airtime the following Tuesday. This coverage, which many journalists questioned, triggered a media chain reaction that probably gave the notorious Dr. Seed more attention than he ever deserved. Even the White House took note, issuing the statement that "The scientific community ought to make it clear to Dr. Seed . . . that he has elected to become irresponsible, unethical, unprofessional should he pursue the course that he has outlined today."

AFTER DOLLY

It was inevitable that other animal species would be cloned, for such work had been in progress at the time of the Roslin Institute's 1996 and 1997 sheep cloning reports. But the public would get in on cloning attempts too. The claims tended to fall into two categories: "have done" and "will do." The "have dones" are the research projects that are completed, written up, peer reviewed, and published in a respected scientific journal, appearing in said journal before hitting the airwaves and tabloids. The "will do" claims tend to skip the scientific publishing labyrinth, instead being announced with great fanfare on TV talk shows or in the newspapers. These are generally interesting ideas that often involve sexy cloning candidates such as woolly mammoths or frozen ice men. But a little digging reveals that the work is speculative at best.

Consider the hoopla over attempts to clone a woolly mammoth. A cover story in the April 1999 issue of *Discover* magazine proclaimed in tiny print "The Remarkable Attempt to" and then in huge print "Clone the Woolly Mammoth." A quick newstand glimpse of the magazine gives the impression that the work is already under way, but the actual article detailed a failed attempt to locate an extractable mammoth. Months later, a story in the Associated Press and a national televised interview with Northern Arizona University geology professor Larry D. Agenbroad discussed a multinational project to unearth (un-ice, actually) a mammoth from an ice field in Siberia, discovered in 1997 by a 9-year-old boy and named Jarkov after him. The mammoth has been flown to Khatanga, where it occupies an ice tunnel where scientists from various nations will pick at its parts, searching for a nucleus that hasn't been shredded to smithereens and whose DNA is intact enough to survive transfer to an enucleated Asian elephant egg. When asked why one would want to do this, Dr. Agenbroad responded, "Why not?" But reconstituted mammoths won't be in zoos anytime soon.

The New York Times ran the Associated Press story, ushering in the familiar "if it's in the *Times* it must be true" public reaction. The "news" about the mammoth cloning then spread everywhere, given so much attention that some people probably thought *The Flintstones* were already upon us. Unfortunately, the media blitz surrounding the mammoth effort may detract from the more realistic possibility of using cloning to save endangered species. Some of that is already reality. "Researchers in New Zealand cloned a weird cow from a distant island. It was the only one left, and there was no way to propagate it. The cow had high salt tolerance, but it wasn't an economically important animal. But this is a way to use cloning to propagate a type in the future—they got 20 clones from that one cow, and this will help to maintain the breed," says Steve Stice.

Also in the "will do" category but definitely a bona fide scientific effort despite the media attention, is the Missyplicity Project at Texas A&M University (Figure 7.5). A wealthy couple donated $2.3 million to clone their dog Missy, a border collie of mixed lineage. Considering the conservative public funding climate where cloning is involved, accepting private funds is certainly understandable. And people are lining up to adopt the dogs that will result from the project, reports Westhusin. But that, too, won't be anytime soon, for the canine is a challenging cloning candidate because of its 6-months-on, 6-months-off estrus cycle. "We don't have the basic reproductive technology worked out. We don't yet know how to cause superovulation in dogs, or synchronize their cycles or do *in vitro* fertilization or *in vitro* culture," Westhusin adds. Cloning Missy will yield some dogs and provide valuable information. "My intent is to learn as much about reproductive physiology as possible, so that we can develop better birth control. The primary focus is to learn reproductive physiology." However, the group will also clone "exceptional individual dogs of high societal value" and "develop relatively low-cost commercial dog-cloning services for the general public" through a service called "Genetic savings and clone."

Oddly, legitimate cloning milestones following Dolly were downplayed. On July 24, 1997, just after Dolly's first birthday, her creators announced the births of five lambs cloned from fetal fibroblasts, four of whom—named Polly, Molly, Holly, and Olly—carry a human clotting factor. Two years later came the Scottish sheep clones Cupid and Diana, harboring a human gene encoding an enzyme used to treat hereditary emphysema. Cloning transgenic animals, even from nuclei from fetal cells, proved that cloning could support transgenesis. Cloning skips a generation and yields one lamb for every 60 nuclear transfers, compared with the 1 in 500 efficiency of microinjecting transgenes.

Bovine cloning progressed too. At the annual meeting of the International Embryo Transfer Society in Boston in mid-January 1998, Robl and Stice announced the births of George and

FIGURE 7.5

The cloning of a dog raises objections among those concerned about the huge numbers of stray animals needing homes. But the donation of several million dollars to fund Missy's cloning also supports basic research at Texas A&M—at a time when government funding for cloning research is unreliable. Other facilities offering future canine clones include Per PETuate in Farmington, CT, and Canine Cryobank in San Marco, CA.
Courtesy of Genetic Savings & Clone, Inc. Copyright 2000 Louis Hawthorne.

Charlie, two transgenic calves cloned from fetal fibroblasts. A feature article in the January 30 issue of *Science* described the not-yet-published results, and the ensuing aborted media response demonstrated that timing is everything in life. "We had 2 days of very intense public interest, with headlines such as 'moo two.' We were very happy to get the publicity, but it ended the second day because the Monica Lewinsky story broke and took over. It spared us from having to talk on *Meet the Press* about cloning, because all they wanted to talk about was Monica," recalls Stice. The full research report appeared in the May 22 issue of *Science* that year.

The cloned calves George and Charlie, along with a sibling, had an interesting beginning. The recipient cells arrived at the University of Massachusetts laboratory directly from midwestern slaughterhouses, which led to some difficult-to-explain negotiations when Federal Express lost a shipment and Robl and Stice had to file a claim for lost cow eggs. The donor cells, fibroblasts, came from a 55-day-old male fetus. They were cultured through two doubling periods, then given bacterial genes encoding the enzyme β-galactosidase and an antibiotic resistance gene. These two genes enabled the researchers to easily detect transgenic cells—they stain blue and grow in the presence of neomycin. A variation on the Dolly protocol, however, was to maintain the cells in the G_1 phase, a point at which they had not yet replicated their DNA. In contrast to G_0 cells, the fibroblasts in G_1 were still in the main part of the cell cycle and therefore were expected to easily support the rapid cell divisions of a new embryo. They did. But the actual work is technically challenging. "The manipulations are done under the microscope at 200 to 400 times magnification.

TABLE 7.4

Cloning Calves from Fetal Fibroblast Nuclei
276 embryos produced by nuclear transfer
33 blastocysts alive after 1 week in culture
28 blastocysts implanted into 11 surrogates
6 cows conceived, still pregnant by day 40
5 cows still pregnant by day 60
1 cow aborted day 249, with grossly abnormal placenta and huge fetus
4 calves born
1 calf died after 5 days

We had to check the eggs to be sure they were enucleated," says Stice, who finds technicians in video arcades because they have the best hand/eye coordination.

The numbers and weights in the calf cloning experiments were reminiscent of the Granada days (Table 7.4). Of the six gestating calves, one died a month before birth, weighing more than a typical newborn. The animal had an enlarged heart, lungs swollen with fluid, and huge umbilical vessels. The calf that died shortly after birth had a similar appearance, with a pulmonary artery wider than the aorta, and umbilical vessels three times their normal diameters. The placentas of both doomed calves were grossly abnormal. Researchers still do not know what causes this problem—the nuclear transplantation, the culture conditions, or the fact that a male and female genome do not contribute to the first cell of the new organism.

Perhaps the most interesting part of the Massachusetts calf trio is their striking differences in coat color (Figure 7.6). "The *Science* paper has a photo of the calves, who were gestated in different uteri. But notice that they have different markings, different locations of color spots on the legs. This is an environmental effect, showing that everything is not genetically programmed. During development, melanocytes migrate at different rates," explains Stice.

Despite the problems with giant animals, the numbers indicate that a "cash calves" technology is feasible. The George and Charlie experiment refined the estimates on how cloning can ease transgenesis. Obtaining one transgenic calf using microinjection requires 2030 embryos placed into 74 recipients, Stice estimates; with cloning, however, it requires 92 embryos placed into 4 recipients.

THE MOUSE, TAKE 2

In 1998, the well-studied *Mus musculus*—the common house mouse—reentered the cloning stakes, becoming the second mammal to be cloned from the nucleus of an adult cell. This feat hit very close to home for developmental biologists and geneticists. In short, if it could be done in a mouse, it could be done in a human.

There were good reasons why McGrath and Solter couldn't clone mice 14 years earlier. "Mice are harder to clone than cattle and sheep, because the embryos are smaller, and the embryonic genome turns on earlier," says Stice (Table 7.5). A cattle embryo takes

TABLE 7.5

Stage of Activation of New Genome

Species	Stage
Mouse	Two-celled embryo
Pig	Four-celled embryo
Human	Four-celled embryo
Sheep	Eight-celled embryo
Cattle	Eight-celled embryo

FIGURE 7.6

The different pigment patterns on cloned calves indicate that clones are not exactly identical. Epigenetic factors sculpt their differences, such as migrations of cells in the developing embryo.
Photograph courtesy of and used with permission from the University of Massachusetts.

several days to turn on its new genes; in mice, that happens in less than 24 hours, barely providing enough time for the delicate genetic material to recover from the manipulations and to be reprogrammed to begin development.

It was fitting, perhaps, that Cumulina, the first cloned mouse, made her debut at a press conference in New York City on Gregor Mendel's birthday, July 22, 1998. As the father of

FIGURE 7.7

Three generations of cloned mice. (Top) Original nucleus donor. (Middle and bottom) Second and third generations of cloned mice.
Photograph courtesy of Ricki Lewis.

genetics, Mendel explained genetic variability; Cumulina and her many cloned siblings represent the ultimate in genetic uniformity (Figure 7.7). The press conference coincided with publication of the work in *Nature Biotechnology* in the same issue in which another article presented the long-awaited DNA fingerprinting evidence that vindicated Dolly as not a fluke. Davor Solter again wrote a "News and Views" article, honoring both reports. The famed mouse's "father" was Teruhiko Wakayama, a postdoctoral research associate then in the laboratory of Ryuzo Yanagimachi, a professor of anatomy and reproductive biology at the University of Hawaii. Cumulina was already nearly middle-aged, having been conceived in August 1997. Wakayama had kept the work under wraps until he was certain that the mouse was truly a clone. He has since moved to Rockefeller University.

The mouse report was well received, perhaps because it was midsummer. The scientific community had waited more than a decade for news of successful mouse cloning, and the public had calmed a bit on the matter of cloning since Dolly. "We had heard rumors about these papers for six months, and I was one of the most active skeptics. One problem I had with the work was that I didn't do it! It is an absolutely beautiful paper, with a sufficient number of mice and experiments to draw conclusions," said Robl at the press conference.

The "Honolulu technique" for cloning mice extends the work on sheep. Cumulina's and Dolly's beginnings differ in several ways (Figure 7.8). Dolly's donor nucleus, from a mammary gland cell, was starved in culture to stall it in G_0, and this step was thought to be the secret to their success. But the Hawaii group showed that the situation is more complex by using three cell types in this stage, with widely varying results. Nuclei from cumulus cells,

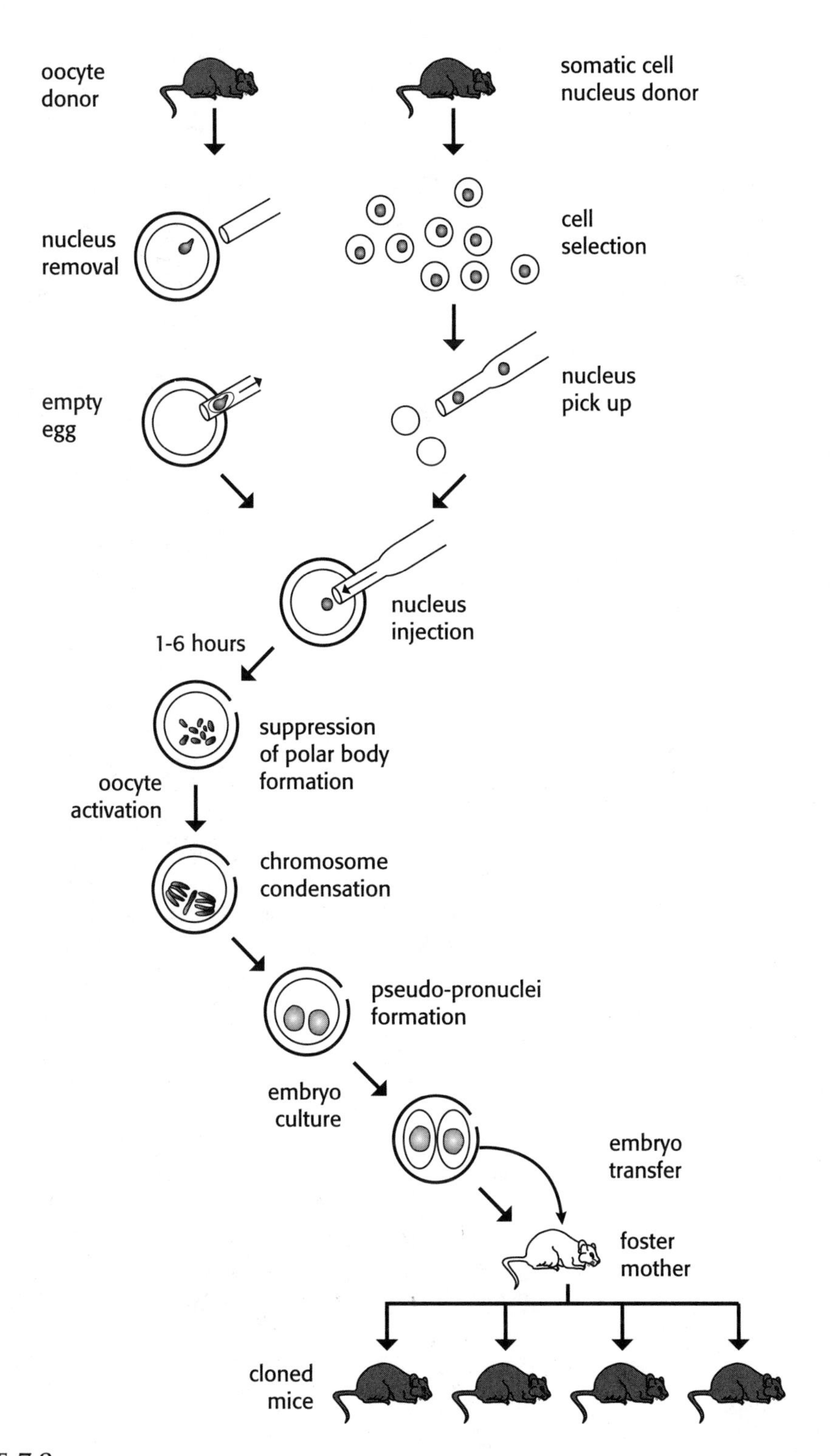

FIGURE 7.8

The Honolulu technique for cloning mice.
Reprinted by permission from Nature *(Wakayama T, Perry ACF, Zuccotti M, et al. Full term development of mice from enucleated oocytes injected with cumuls cell nuclei.* Nature *1998;394:369–374), copyright 1998 Macmillan Magazines Ltd.*

which surround an oocyte, supported mouse development to the blastocyst stage considerably more often than did nuclei from neurons or from the Sertoli cells that nourish developing sperm. These early experiments convinced Wakayama to stick to cumulus cells, and hence the name "Cumulina" for the first success.

In an initial series of nuclear transfers, he placed 142 embryos in 16 foster mothers, yielding 5 live and 5 dead fetuses. A second series transferred 800 embryos to 54 foster mothers, producing 10 healthy pups. The third series was the most visually striking: Agouti (coffee-colored) mice donated the nuclei, black mice provided the enucleated oocytes, and the foster mothers were white mice. As predicted, the cloned mice were agouti, matching the instructions in the donor nucleus. A final series of experiments cloned the clones.

As with many cloning experiments, the cloners aren't quite certain exactly why and how they succeeded in this technique that seems as much art as it is science. Yanagimachi suggests that their quick and clean microinjection technique and delaying egg activation may have made the difference. One to six hours elapsed between nuclear transfer and chemical activation with strontium, which jump-starts development. "They are like sleeping cells that must wake up. During the one to six hours it takes for chromosome condensation, something important happens," says Yanagimachi. That something might be the cell adjusting to the absence of conventional fertilization.

CALVES, REVISITED

The media reports following the mouse announcement paled next to the fuss that embraced Dolly. *The Today Show,* for example, considered the antics of the aforementioned Monica Lewinsky, the deaths of an astronaut and an actor, an ongoing heat wave, the latest lottery news, and a film debut to be more important than the cloning of a mouse. Cumulina quickly and quietly faded from the scene. And cloning announcements to come provoked only brief flutters of news reports, such as the limited media response to a December 11, 1998, article in *Science* detailing the cloning of calves from adult cell nuclei in Japan.

Yukio Tsunoda, a professor of animal reproduction at Kinki University in Nara, Japan, had been perfecting the same technique that led to Dolly since the mid-1980s, and so the Roslin Institute's announcement galvanized him to speed up his work on calves. Although Tsunoda's group was the first to clone cattle from adult cell nuclei, at the time of their publication, five research teams from Japan had birthed 19 such calves, with several dozen more gestating in surrogates. The reason for Japan's bovine boom was a government that has fully supported agricultural biotechnology. The Agriculture Ministry had ruled against cloning from adult cell nuclei only because the consensus was that it couldn't be done. With the birth of Dolly, they lifted the restriction. Previous work with cloning from embryonic cell nuclei had yielded about 400 calves, but the problem of gigantism resurfaced. The average weight of a calf cloned from embryo cell nuclei was 71 pounds, compared with the normal birthweight of 60 pounds. One calf (Figure 7.9) cloned from a fetal cell nucleus set a record at 115 pounds!

Against that discouraging backdrop, Tsunoda's group nevertheless produced eight calves from adult cell nuclei, although four died soon after birth. The researchers starved the donor cells into entering G_0, transferred the nuclei to enucleated oocytes, and activated development with an electrical shock. They used cumulus cells and cells from the oviduct of a cow. The calves were named for their cell source: OVI-1, OVI-2, and CUM-3

FIGURE 7.9

Cloned cows.
Reprinted by permission from Science *(Yoko Kato et al. Eight calves cloned from somatic cells of single adult.* Science *1998;282:1975).*

through CUM-8. The four who died seemed to not fare well in the environment—CUM-3 had pneumonia from heatstroke, CUM-4 and CUM-5 swallowed too much amniotic fluid, and OVI-2 succumbed to abnormal birth and labor. The other research groups currently gestating cloned cattle will provide even more information on the developmental potentials of cells by using donor cells not involved in reproduction, including those from heart, liver, muscle, kidney, and skin.

CLONED OTHERS

In May 1999, the cloned ovines, bovines, and mice were joined by caprines—goats. Researchers at Genzyme Transgenics Corp. in Framingham, Massachusetts, Tufts University, and Louisiana State University reported cloning three goats from the nuclei of fetal fibroblasts, calling them "nuclear-transfer derived animals" instead of the less precise "clones." Goats may be even better candidates for cloning than their predecessors. Compared with cows, their generation time is shorter and their milk production two- to threefold greater, and they very rarely get scrapie (see Chapter 4) or any other spongiform encephalopathy.

Of course, by now cloning from fetal cell nuclei wasn't news—after all, Granada and others had made a business of it—but these she-goats three were different: Each of their cells harbored the human gene encoding antithrombin III, a potential clot-busting drug. Two goat clones hailing from Nexla Biotechnologies in Montreal, Peter and Webster, are transgenic for spider silk! Still, hurdles will have to be surmounted before goat cloning becomes commercially viable, including, again, the high incidence of neonatal deaths, the fact that the donor cells do not live long enough in culture for some manipulations, and the high price tag of $100,000 to $300,000 per cloned transgenic goat. Yann Echelard, of the Genzyme group, notes that automating the procedure, which might ultimately bring down costs, is complex. "The most difficult part is not the nuclear transfer itself, it's all in the surrounding manipulations—maintaining cell lines at the proper stage and age; collecting and selecting oocytes, which is particularly difficult with goats; and transferring and maintaining pregnancies in a herd."

In September 1999, Westhusin and colleague Jonathan Hill from Texas A&M University announced that they had cloned a baby bull, Second Chance, from a 21-year-old prized Brahman named Chance. The elder Chance was no ordinary bull—the unusually even-tempered beast had frequented *The Late Show with David Letterman* and had starred in TV commercials. "This was a companion animal. He was his owners' friend, and part of their livelihood," says Hill, a veterinarian. But like other cloning endeavors, it wasn't easy. 189 nuclear transfers led to six pregnancies, two removed early to study development, and three more dying in utero. Only Second Chance was liveborn. Chance died a few months before Second Chance began, shortly after donating the skin fibroblasts from his belly that provided the genetic instructions for his replica. He could not breed naturally because his testicles had been removed. In march 2000, PPL therapeutics unveiled a quintet of cloned piglets, gestated using an undisclosed new variation on the cloning theme. Pigs have been difficult to clone because their blastocysts are fragile, and at least four must be carried at a time.

Dolly, meanwhile, is doing well. In early spring 1998 she gave birth to Bonnie, following a natural mating with David, a Welsh mountain ram. "She's not a clone, she's just a sheep," says Alan Colman of Bonnie. On March 24, 1999, Dolly bore triplets—two males and a female, again with David. Said Harry Griffen, assistant director of the Roslin Institute in announcing the births, "The birth of Bonnie almost exactly 12 months ago confirmed that despite Dolly's unusual origins, she is able to breed normally and produce healthy offspring. The birth of these three lambs is a further demonstration of this."

HUMAN CLONING: WHEN AND WHY?

What will the future bring? It seems like it is only a matter of time until humans are cloned. "I believe that human cloning will be possible, and will be carried out unless expressly forbidden by society," says Marjori Matzke, head of the department of plant molecular genetics at the Institute of Molecular Biology at the Austrian Academy of Sciences in Salzburg. But attempts may precede results for an unknown time. "Since good sense and responsible behavior are seldom effective barriers to reckless ideas, I expect that the cloning of a human being will be tried soon. But I would not expect…cloning using the Dolly method to be successful for some time," says Harold Shapiro, president of Princeton University and chair of the National Bioethics Advisory Committee.

Biologists differ on just when they realized that human cloning had jumped much closer to reality. For Davor Solter, it was with the births of Megan and Morag. "The moment I read about Megan and Morag, I knew that that was it. We all thought that embryos had to be nuclei donors. It's quite obvious now that that was wrong. A fetal fibroblast is just as differentiated as any adult cell. From then on it was clear that you could clone humans." The cloning of mice from adult cell nuclei confirmed for many biologists that human cloning was not only a possibility, but an inevitability. "It is now clear that cloning is not specific to a single species, and sheep and mice are about as far apart as you can get among placental mammals. Sheep need their embryonic genome by the eight-cell stage, human embryos start using theirs by the four-cell stage, and mice do this by the two-cell stage. There was worry that if reprogramming DNA took a long time, mice and humans would be unable to use their embryonic genomes. So there is now no longer a reason to expect that humans couldn't be cloned as well," says Lee Silver, professor of molecular biology at Princeton University. But not all mammals can be cloned with the same ease. And not all researchers think that human cloning will become reality.

"In my opinion, there won't be clones of humans walking down the street. For one, there are technical difficulties, and the problem of getting people to donate eggs. I don't think society as a whole will ever allow this to happen," says Stice. But it may not be up to either the scientific community or society, as the strange tale of Dr. Seed illustrates. "There is no doubt in my mind that somebody is planning to clone themselves, now. I see a super-rich egomaniac being the first to do it," says George M. Malacinski, professor of biology at Indiana University in Bloomington.

Even if cloning can be regulated the problems of large offspring and their inability to cope with the environment must still be tackled. We may need to look to plants, which have been cloned for many years, for clues as to what may go wrong, according to Matzke. "Problems in cloned organisms might not become visible for years. This is the case with some cloned oil palms, which appear normal until they begin flowering approximately 15–20 years after they have been produced. A recurring defect in the development of female floral parts prevents the production of normal flowers and, hence, oil. The palm oil industry is still trying to find a solution to this problem."

While researchers continue to refine cloning technology, working out the glitches in the "how," many people are thinking about the "why." Looking to the history of cloning, however, does not offer easy answers. Cloning began as a tool to study early animal development. Certainly one needn't clone a human to study how a vertebrate develops—frogs, mice, and other model organisms have done this quite well. But there are other reasons that people may have to clone humans, and they seem to lie more in the realm of science fiction than science fact.

In the novel *The Experiment* by John Darnton, 30-year-old protagonist Jude meets his 25-year-old self, a clone named Skyler who has escaped from a gang of hippie scientists who cloned their children, gestated them five years later, and raised them on an isolated island to provide, if needed, spare parts for their older counterparts. In the spectacular ending, the hippie queen/head scientist gives birth to herself, and the new she is a monster. The overt message is that cloning, after all, cannot circumvent congenital developmental anomalies. The subtle message is that Jude and Skyler are clones, but are vastly different in temperament, personality, and intelligence.

The premise of *The Experiment* isn't necessary, for clones won't be needed for spare parts. Another science fiction writer, Robin Cook, earlier described an alternative in his novel *Chromosome Six*, in which bonobos transgenic for the HLA complex of genes that determines tissue rejection reactions are raised to provide replacement parts for wealthy individuals, also on a remote island. Cook's and Darnton's inspiration may have come from reports of cloning cow embryos that produce dopamine and could cure Parkinson's disease when the embryo tissue is transplanted into the brains of stricken rats, and of PPL Therapeutics' cloning of pigs that lack antigens that could provoke a human immune response. Such pig parts may provide transplants for humans.

Another reason not to clone humans is that, for myriad technical as well as obvious reasons, a clone is not absolutely identical to the original. Thus, cloning a departed loved one or a special individual such as Missy the dog, Chance the bull, or even Jarkov the frozen mammoth doesn't produce the desired exact replacement. And of course cloning to satisfy one's ego is senseless: A clone is a separate individual.

What exactly is it about human cloning that people fear so much? We are not inordinately afraid of twins or triplets—quite the opposite. When multiples occur unnaturally, such as by *in vitro* fertilization, the adorable results end up on tabloid covers and on morning television news shows, gleaning generous donations of diapers, cars, and college tuition.

Manipulating ova and early embryos isn't new enough to provoke fear, although when the first "test tube baby," Louise Joy Brown, was gestated in 1980, her arrival was greeted with the expectation that she might have three heads or bear a Pyrex logo embedded in her thigh. Today the thousands of youngsters born from *in vitro* fertilization are regarded as any other children. Identical fertilized ova have even been separated in time and then implanted in their mother, leading to a few cases of twins born years apart, not unlike Jude and Skyler of *The Experiment.*

CLONING COMES FULL CIRCLE

While we continue to debate and perhaps exaggerate the dangers of cloning, the field has, in a sense, come full circle. The continuing survival of various cloned mammals is addressing long-pondered questions about basic development. "What happens to a genome as an embryo develops from one cell to millions? How are molecular signals sent and interpreted? How is this reversed? What are the biological and technical limitations to cloning? Can all types of cells be reprogrammed? Must all the genes be reprogrammed? Can some differentiated features be retained and cloning still work?" asks Columbia University's Papaioannou.

The very success of cloning answered Weismann's and Spemann's questioning of the irreversibility of a cell's potential posed so many years ago. Yes, a genome can be turned back on. Cloning from various cell sources is telling us how that ability may differ among cell types. Clones are revealing the roles and interactions of nuclei and cytoplasms, the contributions of genes from two parents, and the importance of cell cycle synchrony.

Ironically, what clones are teaching us about basic biology is coming not from their uniformity, but from their differences. For clones are *not* exact replicas, for a variety of reasons:

- A nucleus from a somatic cell may be structurally the same as one that results from the merger of sperm and egg, but it is not functionally equivalent. The "imprint" of passing through the germline is not stamped on the starting cell of a clone, with as yet unknown consequences. For some genes, one copy is turned off as a new individual forms, depending on which parent transmits it. Absence of this imprinting in cloned animals might have an effect that would not be noticed for several generations. "Modifications such as DNA methylation are required to silence genes that are not needed in given cell types as development unfolds in plants and animals. If these epigenetic modifications are not fully erased in somatic nuclei that are used for cloning, there will be problems in activating the affected genes at the appropriate stage of development," says Matzke. The result, she adds, would be genetic clones that nonetheless do not look, or function, alike.
- Cells reveal their ages in the lengths of their chromosome tips, or telomeres (see Chapter 5). The older the cell, the shorter the telomeres. In the donor

nucleus that became Dolly, the telomeres were shorter than those in a fertilized ovum because they spent six years in a sheep. As a result, Dolly's telomeres are much shorter than they should be for her age from birth, although she does not appear to be aging prematurely. However, telomeres in cloned cows and mice are *longer* than normal.

- DNA from adult donor cells has had years to accumulate mutations. Such a somatic mutation might not be noticeable in one of millions of mammary gland cells, but it could be devastating if that somatic cell is used to start an entire new individual. In fact, the cloning of carrots and other plants in the 1970s and 1980s revealed this problem: Researchers would culture plants from genetically identical somatic cells only to find heritable differences arise, such as yellow leaves or sweeter fruits. These "somaclonal variants" provide interesting new plants. A similar phenomenon in a cloned animal could be disastrous. Certain somatic mutations in animals, for example, cause cancer.
- At a certain time in early prenatal development in all female mammals, one X chromosome is inactivated in each cell. Whether the turned-off X is from the mother or the father occurs at random in each cell, creating an overall mosaic pattern of expression for genes on the X chromosome. It would be very unlikely, indeed nearly impossible, for the pattern of X inactivation in Dolly to match that of her nucleus donor, the 6-year-old ewe.
- Mitochondria, the cellular organelles that extract energy from nutrients, contain DNA, albeit only a few dozen genes. Dolly's mitochondria are descended from those in the recipient oocyte's cytoplasm, not from the donor cell. Genetically speaking, there is a little bit of the recipient in Dolly.

More obvious, perhaps, than these precise, biochemical reasons why a clone isn't really a clone is another powerful factor—the environment. The coats of cloned calves form different color patterns because of cell movements. Another bovine example of the effect of the environment on clones is highly prized Matsuzaka beef, which sells for about $100 a pound. Cloning a herd of the valuable beasts might seem a wise investment, until one considers just how their elite muscle tissue forms—from a steady diet of beer, and daily massages that evenly distribute the flecks of fat that make the meat so tender. The "trait" of tender meat comes after the genetic program is set and activated.

Clones are often compared to identical twins. However, twins are actually more alike than are clones, who may not share a prenatal environment and aren't subjected to the strictly human psychological effects of being dressed alike and gawked at. Still, anyone who knows identical twins knows too that they are not the same in every way. Experiences, nutrition, stress, exposure to infectious diseases, and other environmental influences also mold who we are.

On a global level, cloning seems the very antithesis of the effort to preserve biodiversity. On a personal level, it is simply undesirable to many. Steve Stice, accused of personally cloning while his wife carried twins (they turned out to be fraternal) sums it up well: "I wouldn't want clones. Individuality is something to be cherished."

REFERENCES

Aldhous, Peter. "Cloning's owners go to war." *Nature* 405:610–612 (2000).

Briggs, R., and T. J. King. "Transplantation of living nuclei from blastula cells into enucleated frog eggs." *Proceedings of the National Academy of Sciences* 38:455–463 (1952).

Campbell, K. H. S., et al. "Sheep cloned by nuclear transfer from a cultured cell line." *Nature* 380:64–67 (1996).

Chan, A. W. S., et al. "Clonal propagation of primate offspring by embryo splitting." *Science* 287:316–321 (2000).

Cibelli, J. B., et al. "Cloned transgenic calves produced from nonquiescent fetal fibroblasts." *Science* 280:1256–1258 (1998).

Gurdon, J. B. "The developmental capacity of nuclei taken from intestinal epithelial cells of feeding tadpoles." *Journal of Embryology and Experimental Morphology* 10:622–640 (1962).

Gurdon, J. B., et al. "The developmental capacity of nuclei transplanted from keratinized cells of adult frogs." *Journal of Embryology and Experimental Morphology* 34:93–112 (1975).

Kato, Y., et al. "Eight calves cloned from somatic cells of a single adult." *Science* 282:2095–2098 (1998).

McGrath, J., and D. Solter. "Nuclear transplantation in the mouse embryo by microsurgery and cell fusion." *Science* 220:1300–1302 (1983).

McGrath, J., and D. Solter. "Inability of mouse blastomere nuclei transferred to enucleate zygotes to support development *in vitro*." *Science* 226:1317–1319 (1984).

McLaren, Anne. "Cloning: pathways to a pluripotent future." *Science* 288:1775–1780 (2000).

Schnieke, A. E., et al. "Human factor IX transgenic sheep produced by transfer of nuclei from transfected fetal fibroblasts," *Science* 278:2130–2133 (1997).

Solter, D. "Dolly is a clone—and no longer alone." *Nature* 394:315–316 (1998).

Vogel, Gretchen. "In contrast to Dolly, cloning resets telomere clock in cattle." *Science* 288:586–587 (2000).

Wakayama, T., et al. "Mice cloned from adult cell nuclei." *Nature Biotechnology* 394:369–373 (1998).

Willadsen, S. M. "A method for culture of micromanipulated sheep embryos and its use to produce monozygotic twins." *Nature* 277:298–300 (1979).

Willadsen, S. M. "Nuclear transplantation in sheep embryos." *Nature* 320:63–65 (1986).

Wilmut, I., et al. "Viable offspring derived from fetal and adult mammalian cells." *Nature* 385:810–813 (1997).

CHAPTER 8

HOMOCYSTEINE AND HEART HEALTH

> Nature is nowhere accustomed more openly to display her secret mysteries than in cases where she shows traces of her workings apart from the beaten path; nor is there any better way to advance the proper practice of medicine than to give our minds to the discovery of the usual law of Nature by careful investigation of cases of rarer forms of disease. For it has been found, in almost all things, that what they contain of useful or applicable nature is hardly perceived unless we are deprived of them, or they become deranged in some way.
>
> —William Harvey, 1657

By observing the abnormal, we can infer the normal. So wrote William Harvey (1578–1657), the English physician best known for deciphering the pathway of blood in the circulatory system. Many times in the history of medicine, people with rare "inborn errors of metabolism" have unwittingly supplied clues that helped researchers piece together biochemical pathways, with the resulting insights eventually applied more broadly. This was the case for the discovery of how lipoprotein molecules carry cholesterol in the circulation, and for the identification of another biochemical that affects cardiovascular health but is not nearly as familiar, an amino acid called homocysteine.

OF LDL, HDL, AND HOMOCYSTEINE

Children with familial hypercholesterolemia (FH) are one in a million, inheriting two mutant genes from each carrier parent. The children's blood vessels become so packed with cholesterol that yellow deposits are visible and palpable beneath the skin, particularly in the crooks of the elbows and behind the knees. At a microscopic level, their liver cells are denuded of low density lipoprotein (LDL) receptors. Normally these receptors bring cholesterol from the bloodstream inside the liver cell, where it signals the cell to cease its own manufacture of the substance. But when the incoming LDL particles can't dock, the message is never received, and the liver continues to make cholesterol—despite large amounts already in the circulation from the diet. Cholesterol continues to build up in the blood, and sooner or later heart and blood vessel disease result—sooner in the youngsters, later in their parents, who have half the normal number of LDL receptors.

Studying children with FH helped Michael Brown and Joseph Goldstein, at the Southwestern University School of Medicine in Dallas, identify LDL receptors and learn how the liver manufactures and regulates cholesterol, for which they won the 1985 Nobel Prize in physiology or medicine. Their work led to the development of cholesterol-lowering drugs that today help millions keep their levels down—and bring in some $50 billion a year to pharmaceutical companies.

A different type of inborn error of metabolism targets high density lipoproteins (HDL). Whereas LDL particles are packed with cholesterol, HDL particles are mostly protein, and so a low LDL and high HDL cholesterol profile signals heart health. People with Tangier disease have a greatly increased risk of cardiovascular disease, although for a different reason than the children with FH: They have low levels of a type of HDL in the blood. The disease is mostly found on Tangier Island, in the Chesapeake Bay. The people who settled the island in 1686 remained there, isolated from the mainland and marrying among themselves, thus conserving the gene and its associated disease. In 1999, researchers identified the mutation that causes Tangier disease. Understanding how HDL plummets in these people, just as understanding how LDL accumulates in children with FH, is likely to reveal more about the workings of the heart and blood vessels, and lead to development of new types of cholesterol-lowering drugs.

Physicians typically determine a patient's LDL and HDL levels and use this information to estimate the risk of developing cardiovascular disease. The knowledge that a healthy heart profile includes low LDL and high HDL came both from studying people with rare inborn errors and from large-scale clinical trials that traced LDL and HDL cholesterol levels over many years in many individuals. Children with inborn errors of metabolism have also led biochemical sleuths to identify another independent risk factor for heart and blood vessel disease, the amino acid homocysteine.

The unfolding story of discovery of the link between elevated homocysteine levels in the blood and damage to arteries goes straight to the heart of scientific inquiry. It is a tale of a hypothesis evolving into a theory, and of many researchers testing that theory, posing and answering questions, and probing possibilities. Over time and with accumulating data, the association has become a correlation, and then the correlation a cause, which will perhaps culminate with understanding of an underlying disease mechanism. And like the stories of prions and archaea, of stem cells and cloning, the saga of the discovery of homocysteine's role in cardiovascular health and disease also has elements beyond science—elements of economic motivation, of the human reticence to accept an unexpected or unpopular idea, and of competition and professional jealousy.

CYCLES AND PATHWAYS

Beginning biology students typically pale upon encountering "intermediary metabolism," the intricate and interlocking cycles and pathways that connect the hundreds of different biochemicals that run a cell. At the simplest level, an enzyme catalyzes a reaction that transforms a starting material (substrate) into a product by lowering the amount of required energy. Enzyme-catalyzed reactions are linked into chains, as the products of some become the substrates of others, building the pathways of metabolism. Sometimes there is a practical payoff to deciphering a biochemical pathway or even part of one—understanding how a certain set of symptoms arises, so that new treatments can be more targeted to the biological cause.

An inborn error of metabolism results from a missing or malfunctioning enzyme. The glitch sets into motion a chemical cascade of sorts, a chain reaction of shifting levels of biochemicals. If that enzyme is part of a pathway, then the block causes levels of the product to drop as levels of the substrate rise. Like a dam that exerts effects both upstream and downstream, compounds further back in the pathway accumulate, while compounds that would normally result from further reactions of the product diminish (Figure 8.1). In impaired metabolism, any of the chemicals whose levels veer from normalcy can affect health. Sometimes defects in different enzymes of a pathway cause the same symptoms, such as the inability of the blood to clot. The various forms of hemophilia, for example,

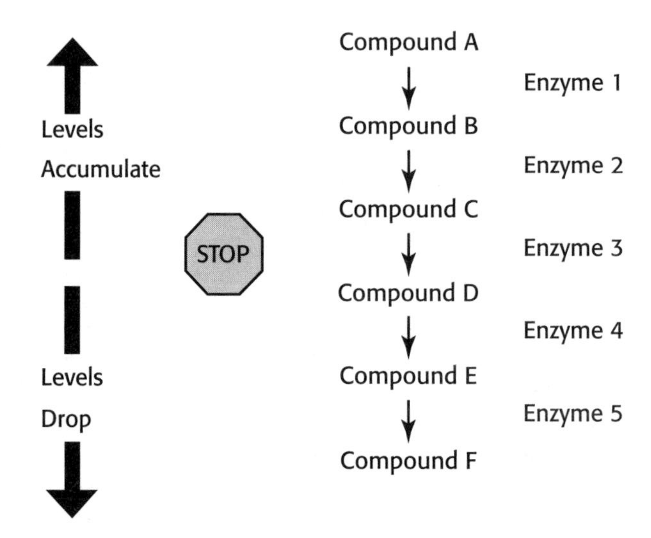

FIGURE 8.1

A block in a biochemical pathway causes compounds before it in the sequence of reactions to accumulate, and levels of the compounds synthesized after the block to fall.

stem from mutations in the genes encoding any of 11 different enzymes that take part in the blood clotting pathway.

English physician Archibald Garrod was the first to note the connection between enzyme deficiencies and inherited disease. In 1902 he described alkaptonuria, a disorder that darkens urine and the ear tips and causes arthritis. The disease was discovered in mice when the wood shavings in their cages turned black when the animals urinated! Lack of the enzyme causes homogentisic acid to build up, which is excreted in the urine and causes the symptoms.

Figuring out the biochemical pathway for the production of homocysteine in the body was difficult for several reasons. First, the route to synthesis is complex, including both a straight-line section of sequential, enzyme-catalyzed reactions and a cycle. How these two parts fit together was not, for a while, obvious. Second, the amount of this amino acid in the bloodstream is vanishingly small. Biochemists simply didn't know it was there, because homocysteine is not one of the 20 amino acids that are found in food and are part of the body's protein molecules. The inability to detect low blood and urine levels of homocysteine delayed its association with health for many years.

Homocysteine was identified by Vincent DuVigneaud, a biochemist at George Washington University in the late 1930s and then at Cornell University Medical College, where he was when he won the Nobel Prize in chemistry in 1955 for synthesizing the first peptide hormone, oxytocin. His discovery of homocysteine began with an investigation of methionine, which is distinctive among the 20 amino acids found in biological proteins for its methyl group (CH_3). A methyl group is a one-carbon compound that is passed from one compound to another, activating different molecules. Methionine is a highly reactive amino acid precisely because of its tendency to lose and gain this appendage, which it readily transfers to other compounds, completing their syntheses. DuVigneaud found that when methionine gives up its methyl, it becomes homocysteine—but that

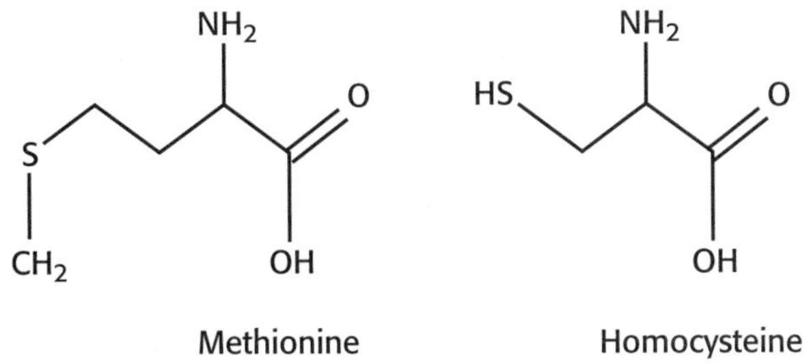

Methionine Homocysteine

FIGURE 8.2

Homocysteine forms when methionine transfers its methyl group (CH_3) to another compound. Homocysteine can be remethylated, which regenerates methionine, but the reaction is not simply the reverse of the reaction that forms homocysteine.

homocysteine usually doesn't accumulate (Figure 8.2). It can either pick up a methyl from an activated form of the vitamin folic acid and thereby regenerate methionine, or it can be dismantled, producing yet another amino acid, cysteine. DuVigneaud uncovered only some of the pieces to this puzzle. Others discovered that vitamin B_{12} is necessary for reforming methionine, and that vitamin B_6 is needed for the straight-chain part of the pathway that transforms homocysteine to cysteine and other compounds.

Another early researcher who studied homocysteine, if indirectly, was Laurence Pilgeram. As a young postdoctoral research associate in biochemistry at the University of California, Berkeley, he took a closer look at the interplay between homocysteine and methionine. In 1954 Pilgeram showed that when the methionine level falls, there aren't enough methyl groups to activate another compound that normally keeps blood from clotting. The body must normally suppress blood clotting, unless a wound requires sealing. As a result of lowered methionine, blood clots form in arteries where and when they aren't needed. The clots either stay in place as thrombi and block blood flow, or they move in the circulation as emboli and lodge elsewhere, possibly causing heart attacks, strokes, or other blockages. In addition, the extra fibrin strands that are laid down in the vessel as a clot forms create an uneven surface that attracts fats, particularly LDL particles. Eventually, deposited cholesterol builds up as lumps called *atheromas*, which means "porridge-like." The atheromas, the hallmark of atherosclerosis, impede blood flow. Homocysteine, according to Pilgeram, might contribute to the development of atherosclerosis because blocking its ability to pick up CH_3 reduces the methionine level, which triggers inappropriate blood clotting.

Pilgeram's thoughts at the time presaged what was to come in the 1960s and 1970s. "I was troubled by the preoccupation of science and medicine with cholesterol as the cause of atherosclerosis, when autopsies revealed a thrombus or embolus as the cause of death. It set me to wondering why people didn't study blood clotting," Pilgeram recalls. His findings on methionine deficiency were important enough to have been published in *Science* in 1954. Pilgeram, still a beginning scientist, added the name of his sponsor, David Greenberg, to the paper as a courtesy. When Greenberg was invited to present the work at a prestigious scientific meeting in 1955, he deferred to Pilgeram, saying that he and DuVigneaud were the only ones in the world who, at the time, knew anything about the interplay between the two amino acids. Pilgeram gave the talk, his legs shaking in fear as he sat among a panel of very senior scientists, but he did well.

Pilgeram's "unifying theory of atherogenesis" wed blood clotting, cholesterol buildup, and fluctuating levels of methionine and homocysteine in contributing to heart disease. But homocysteine would turn out to have a more direct role in cardiovascular health, and assembling the parts of the pathway would be necessary to show this. That goal fell to a team at the National Institutes of Health (NIH) hospitals in Bethesda, with the help of an 8-year-old mentally retarded child named Connie.

CLUES FROM SICK CHILDREN

Between 1962 and 1965, scattered reports of children with large amounts of homocysteine in their urine (an inborn error of metabolism called homocystinuria) appeared in the medical literature. Before 1962 the analytical tools to test for this amino acid didn't exist, but it is likely that physicians had noted children with the constellation of signs and symptoms of homocystinuria. These can include mild to moderate mental retardation, light hair, a ruddy complexion, dislocated lenses, and dangerous blood clot formation.

The first case of homocystinuria was identified in Belfast, but other affected children were being studied in Madison, Wisconsin, and in Philadelphia. George Spaeth and G. Winston Barber, at the Wills Eye Hospital in Philadelphia, referred Connie to a group at the NIH led by Len Laster for further analysis. Not many people at the time knew anything at all about homocysteine, but Laster quickly brought on board the few who did: Harvey Mudd, James Finkelstein, and Fil Irreverre. This was a time when biochemists were physicians, talented individuals who could care for sick children as well as analyze the contents of test tubes. Today's biochemists tend to be more versed in molecular biology than medicine.

Because Connie had high blood levels of both homocysteine and methionine, the researchers suspected that her blockage was beyond the point in the pathway where homocysteine recycles back to methionine. Specifically, she might lack the enzyme that normally catalyzes the reaction of homocysteine to form cysteine, the straight-line part of the complex metabolic pathway. That enzyme is cystathionine β-synthase, or CBS. Connie's high levels of methionine indicated that homocysteine could still pick up methyl groups. The fact that homocysteine was being recycled to methionine even when methionine was abundant meant that this reaction must not be simply the reverse of the one that forms homocysteine to begin with. Homocysteine must regenerate methionine some other way (Figure 8.3).

To demonstrate that Connie's symptoms were the result of CBS deficiency, the researchers had to show that her cells couldn't synthesize the enzyme. But not all tissues make CBS, and the elusive enzyme couldn't be detected in easily accessible tissues such as skin or blood. The researchers needed cells from Connie's liver, where the enzyme is produced. Jim Finkelstein, today chief of biochemical research at the Veterans Administration Medical Center in Washington, DC, will never forget the experience of acquiring the tissue. As a gastroenterologist he had taken biopsies from children before, but he was especially concerned that Connie might be extremely frightened because she was mentally retarded. "It was December, 1963, and I was responsible for 8-year-old Connie. We did dress rehearsals with a needle. It was a nightmare. I did the biopsy, knowing that if she moved while I was getting the tissue, it could be a disaster," he recalls. Connie didn't move, and Finkelstein remembers her tears following the procedure.

Connie's liver cells indeed lacked CBS, and by knowing this, the NIH team could finally connect the two parts of the pathway—the cycle that her cells could complete, and the

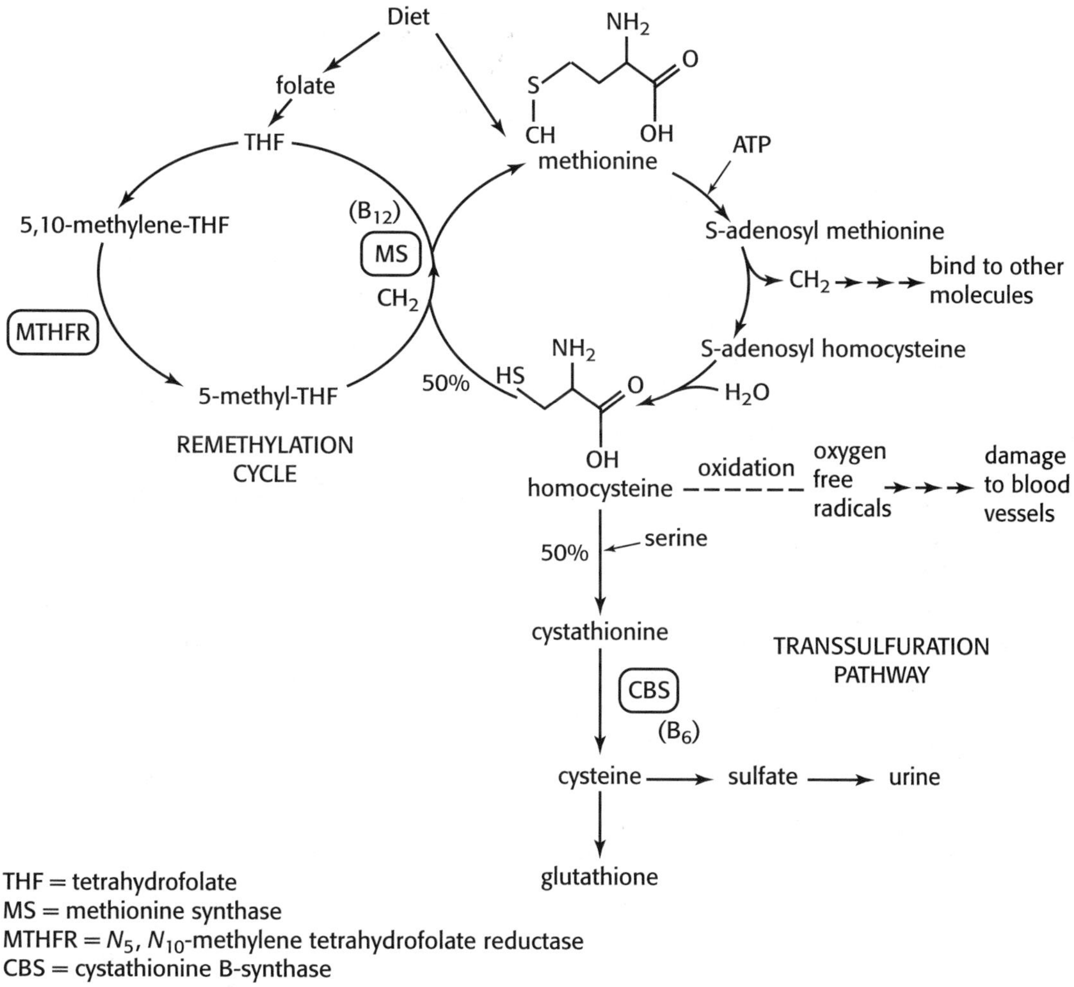

Diet
folate
THF
5,10-methylene-THF
MTHFR
5-methyl-THF
REMETHYLATION CYCLE
(B_{12})
MS
CH_2
50%
NH_2
S
CH
O
OH
methionine
ATP
S-adenosyl methionine
CH_2 → → → bind to other molecules
S-adenosyl homocysteine
H_2O
HS
NH_2
O
OH
homocysteine
oxidation
oxygen free radicals
damage to blood vessels
50%
serine
cystathionine
TRANSSULFURATION PATHWAY
CBS
(B_6)
cysteine → sulfate → urine
glutathione

THF = tetrahydrofolate
MS = methionine synthase
MTHFR = N_5, N_{10}-methylene tetrahydrofolate reductase
CBS = cystathionine B-synthase

FIGURE 8.3

Homocysteine can be remethylated to regenerate methionine. In some cells, such as in the liver, about half of the homocysteine enters the transsulfuration pathway. Along this route, homocysteine combines with another amino acid, serine, to form cystathionine. Next, cystathionine β-synthase (CBS) catalyzes the reaction that breaks apart cystathionine, which leaves the amino acid cysteine. Finally, cysteine picks up a sulfate group and exits the body in the urine, or is converted to glutathione. CBS requires vitamin B_6 to function. Homocysteine may also be oxidized, producing free radicals that can harm artery linings. This may be a direct way that homocysteine damages blood vessels.

chain that they couldn't. The group's 1965 paper in *Science* was the definitive description of classic CBS deficiency homocystinuria. With that information, Spaeth and Barber in Philadelphia were able to develop vitamin B_6 therapy, which helped young Connie and some of her relatives, as well as some members of other affected families. Since then, other treatments, also based on understanding the homocysteine pathway worked out at the NIH, have been developed. But back in 1965, the NIH group had not only connected the pieces of DuVigneaud's pathway but had also stamped a name and a face onto it.

A KEY CONNECTION

Because several different enzymes participate in biochemical pathways, a disease such as homocystinuria can arise in several ways. Variants of homocystinuria even rarer than CBS deficiency, which affects 1 in 200,000 individuals, were eventually discovered. These other disorders are caused by defects in methionine synthase, the enzyme that adds the methyl group back onto homocysteine to regenerate methionine, and in methylene tetrahydrofolate reductase (MTHFR), an enzyme that alters folic acid so that it can donate a methyl group. But it was a defect in the metabolism of vitamin B_{12}, which is necessary for methionine synthase to work, that captured the attention of a young pathologist in Boston, Kilmer McCully (Table 8.1).

McCully credits his eclectic background with preparing him to see connections between fields. After completing his medical training, McCully worked at various laboratories investigating amino acid biosynthesis, cholesterol metabolism, and fungal genetics. In 1968, as a pathology resident at Harvard-affiliated Massachusetts General Hospital, McCully heard a presentation by two pediatricians in which they described a 9-year-old girl who was mentally slow and had dislocated lenses. The symptoms rang a bell with one of the pediatricians, who had studied in the Belfast clinic where the first case of homocystinuria had been described in 1962. Another clue came when the child's mother mentioned to the doctors that the girl's uncle had died of something similar at age 8 more than 30 years earlier. The mother remembered that his case had been so unusual that it had been described in a medical journal. McCully, interested in rare diseases, went to look for it and found case 19471 of the November 23, 1933, issue of the *New England Journal of Medicine.*

According to the old journal, the boy had been admitted to the hospital following eight days of vomiting, drowsiness, and headache. However, the fact that he had been born with dislocated lenses, an abnormal hip joint, and mental retardation suggested that he was suffering from more than an acute illness. In the hospital he began to show signs of a stroke, his left side weakening and the reflexes vanishing. He died three days later of stroke. As a pathologist, McCully was intrigued by a disease of the aged in a young child. He decided to investigate further.

Since the boy's case had been published in so prominent a journal, McCully suspected that the slides taken at the time of autopsy might have been saved. They were, as well as

TABLE 8.1

Vitamins Crucial for Homocysteine Metabolism

Vitamin	Associated Enzyme	Function
B_6	Cystathionine β-synthase	Converts cystathionine to cysteine
B_{12}	Methionine synthase	Recycles homocysteine to methionine
Folic acid	Methylene tetrahydrofolate reductase (MTHFR)	Adds methyl group to homocysteine (folate) to regenerate methionine

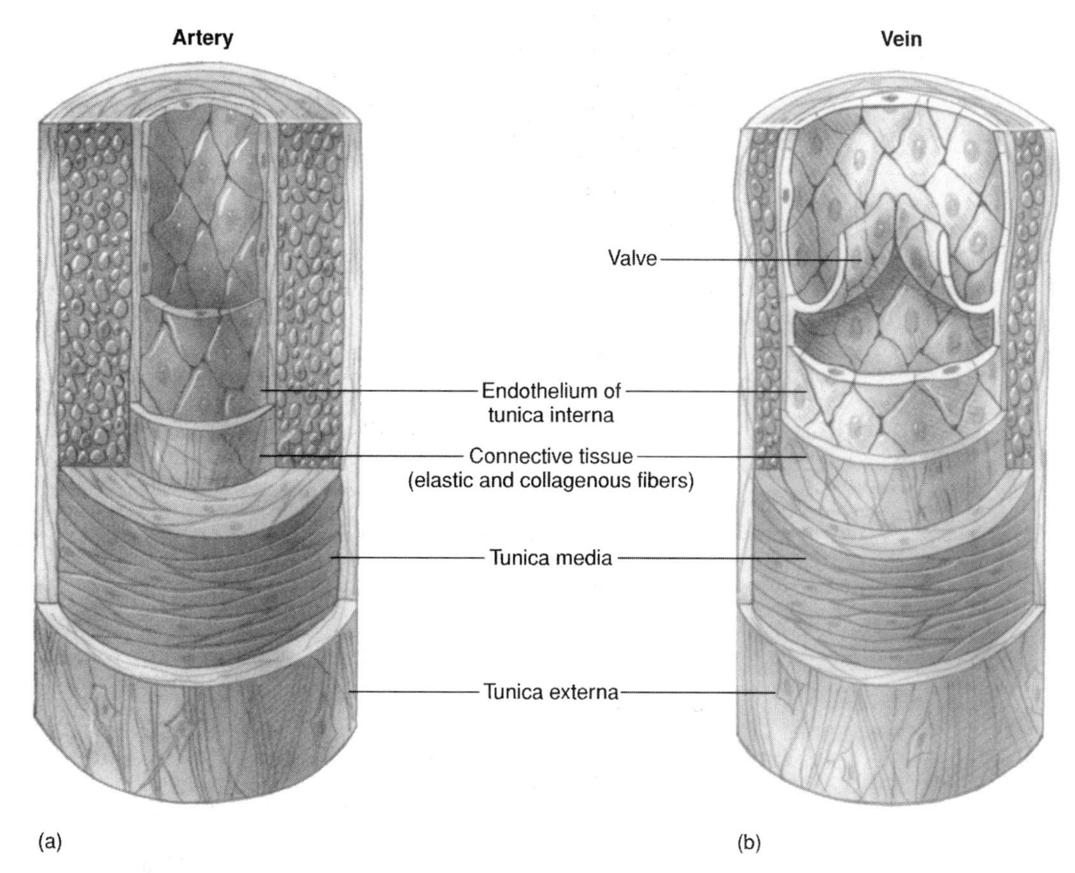

FIGURE 8.4

Blood vessel walls are multilayered. Cardiovascular disease results from a series of insults, including disruption of the smooth inner endothelial lining that can snag lipid-carrying particles, inappropriate fibrin deposition and clot formation, and overgrowth of the smooth muscle layer.
Shier D, et al. Hole's human anatomy and physiology. *7th ed. McGraw-Hill, 1996. Reproduced with permission of The McGraw-Hill Companies.*

paraffin blocks containing preserved tissue, but the wax had melted during years of storage in an attic. McCully had technicians make new slides from the material in the paraffin, and he did his own analysis. He found unmistakable signs of arteriosclerosis, the hardening and narrowing of the arteries that precedes atherosclerosis, the fatty buildup.

McCully confirmed from the slides that stroke had caused the boy's death, and also noted that the carotid artery had the thickened walls and constricted passageway that one might expect in an elderly person who had suffered years of artery disease. A clot had blocked the carotid, contributing to death, and the artery was also obscured with masses of loose fibrous connective tissue—but notably not fatty material. In addition, the usually smooth elastic layer of the artery (Figure 8.4) was frayed and discontinuous. But McCully was astonished at what he found in slides of parts of the body not directly implicated in the boy's death. Throughout the thymus, adrenal glands, kidneys, and heart, the arteries bore microscopic changes: Masses of fibrous tissue hugged the vessel walls as the muscle layer beneath the innermost, usually smooth endothelium overgrew.

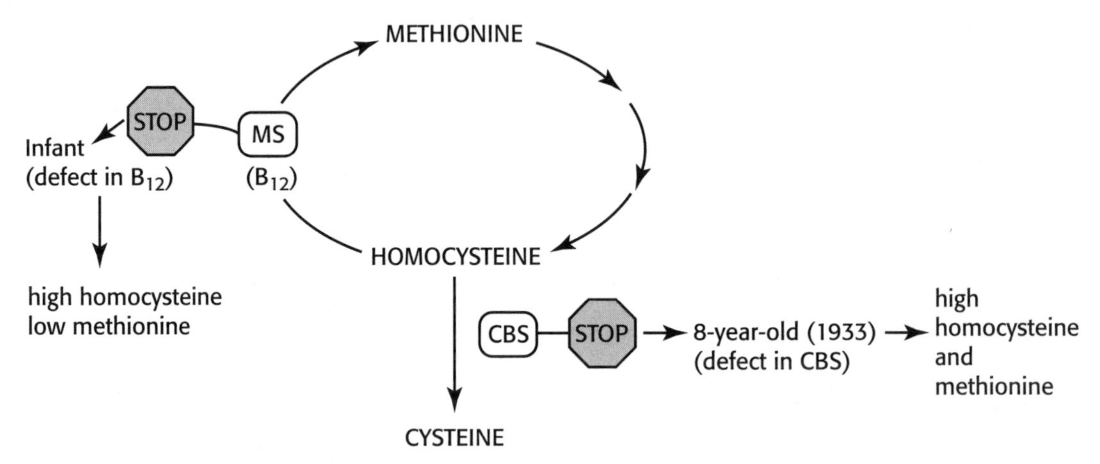

FIGURE 8.5

Studying the levels of homocysteine and methionine in children with different forms of homocystinuria enabled researchers at the National Institutes of Health to connect the two parts of the pathway, components of which Vincent DuVigneaud had identified earlier.

The pathologist who had performed the autopsy in 1933 had also noted vascular changes, but McCully was the first to note the resemblance to arteriosclerosis.

Some months later, McCully heard of a second patient with homocystinuria who had been studied at the NIH, an infant who had other signs of a deranged metabolism. He had high homocysteine levels, but also low levels of methionine and high cystathionine, which is the intermediate compound between homocysteine and cysteine. In contrast, Connie (who McCully knew about from the 1965 *Science* article) and the present girl and her long-deceased uncle had had high methionine levels, because they could recycle homocysteine to methionine. Their blockage, in CBS, occurred in the straight part of the pathway. The infant had a different inborn error, which the NIH team identified as a defect in vitamin B_{12} metabolism. This block affected the functioning of methionine synthase, which requires the vitamin as a cofactor to recycle homocysteine back to methionine. Without B_{12}, the levels of methionine fell. However, the liver could still process homocysteine because CBS worked. Excess homocysteine spilled over into the urine because it couldn't be recycled back to methionine (Figure 8.5).

The baby had been admitted to the hospital because of failure to gain weight, intense vomiting, and lethargy. He suffered convulsions and cardiac arrest, and died in the hospital, not yet 8 weeks old. Once again, McCully examined the autopsy slides, and found that the child's cardiovascular system had detonated. Although the veins and aorta appeared to have been spared, many of the other arteries, of all sizes, were grossly abnormal, swathed in coats of loose fibrous connective tissue, their normally smooth inner linings split and frayed like peeling paint in some places, yet ballooning outward elsewhere. These defects had narrowed the child's left coronary artery enough to cause the cardiac arrest, and if that had not killed him, similar damage in his pulmonary arteries might have had the same result. The boy's arteries bore the marks of destruction in his thymus, stomach, brain, small intestine, heart, thyroid, testes, kidneys, liver, pancreas, and adrenal glands. Like the 1933 case, though, his arteries were noticeably devoid of fatty plaques, and the blockages

TABLE 8.2

Dietary Sources of Methionine

Food	Grams per 100-gram Serving
Brazil nuts	5.8
Eggs	3.2
Cow's milk	2.9
Meat and fish	2.7
Nuts and cereals	1.4–1.8
Human milk	1.4
Most fruits and vegetables	0.9–1.2

were fibrous. There simply might not have been enough time for cholesterol and other fats to have been deposited. In the doomed baby boy, arteriosclerosis killed before atherosclerosis had a chance to develop.

McCully compared the cases. The boy who died in 1933 and the one who had just died in 1968 had cardiovascular systems ruined in apparently similar ways, yet they had different metabolic defects—one had high methionine levels and the other low. What they had in common was excess homocysteine. Therefore, McCully reasoned, could homocysteine be destroying the blood vessels? He knew that animal studies from the 1950s had linked high blood homocysteine concentration and damaged arteries. McCully's observations on the children also seemed to contradict Pilgeram's hypothesis that low methionine level damages blood vessels, by showing that cardiovascular damage occurred whether methionine level was high or low. But he did not cite Pilgeram's 1955 *Science* paper.

In a report published in 1969 in the *American Journal of Pathology,* McCully unveiled what he would eventually call the homocysteine theory of arteriosclerosis. He extrapolated from the children with exceedingly rare inborn errors to heart and blood vessel disease among the general meat-eating, fat-craving public. His idea made sense: If genetically induced homocysteine excess narrowed and hardened the arteries, could a diet-induced excess, although not nearly as extreme, do the same, perhaps at a much slower rate? After all, foods rich in methionine, which could elevate homocysteine level, are the same ones that had long been linked to heart disease (Table 8.2).

The message in McCully's paper was clear—perhaps cholesterol and saturated fats weren't the only causes of the high rates of cardiovascular disease in Western nations. Too much methionine from meats, combined with too little B vitamins from fruits and vegetables, could prime the arteries in a way that eases fat deposition. But his paper didn't attract much, if any, notice. Journalists hardly peruse the *American Journal of Pathology* for story ideas, and even if they had, in July 1969 McCully's ideas would have had to compete for headline space with man's first steps on the moon and half a million young people converging on a muddy farm in upstate New York for the Woodstock festival. Other biomedical researchers regarded the report as a clever connection, but part of a continuum of investigation that had been ongoing for decades.

Scientists are doubters by nature, and soon questions of logic arose concerning the validity of extrapolating from the extremely high levels of blood homocysteine in children

TABLE 8.3

Concentration of Homocysteine in Blood Plasma

Condition	Concentration (μmol/L)
Normal	5–15
Moderately elevated	15–30
Intermediately elevated	>30–100
Severely elevated	>100

with inborn errors to people with the more common, adult-onset forms of cardiovascular disease. The numbers just didn't make sense (Table 8.3). "Half of the kids with CBS deficiency homocystinuria are responsive to pyridoxine [B_6]. You can get their homocysteine levels down to 50 or 80, obviating all the cardiovascular events. It is hard to understand how, if you can get these kids down to 50 and lower their risk, why a healthy person with a level of 15 would show increased risk," asks Jim Finkelstein.

Another possible reason for the lack of attention initially paid to McCully's 1969 paper was the perception of elevated cholesterol as the major risk factor for cardiovascular disease. Even today, a visit to a public library reveals shelves of books on cholesterol, with many a heart health book not even listing "homocysteine" in the index. Nor is homocysteine prominent on the American Heart Association (AHA) website.

At the time of Kilmer McCully's report on a possible link between the little-known homocysteine and heart disease, cholesterol was fast becoming a very big deal. Evidence for a link between dietary fat and clogged arteries had been shown as far back as 1908 in force-fed rabbits, and since then in many other species. Mountains of anecdotal and epidemiologic evidence supported an association between a fatty diet, high serum cholesterol, and risk of heart disease.

As the 1970s dawned, government health agencies were gearing up to bombard the public with anticholesterol messages, as the National Heart, Lung and Blood Institute (NHLBI) began planning its huge Lipid Research Clinics Coronary Primary Prevention Trial. Nearly 10 years and $150 million later, that study would cement cholesterol's reputation as the enemy by linking a 1% decrease in total and LDL cholesterol to a 2% decrease in the incidence and death rate from coronary heart disease. At the same time, major pharmaceutical companies were developing new cholesterol-lowering drugs. "In the 1970s, the cholesterol and lipid theory was favored. A lot of people had invested their careers in this. There was no interest in someone coming along and saying that cholesterol was really secondary, and the homocysteine theory does put cholesterol in second place. I do not think it is the primary cause of artery disease," maintains McCully.

Other homocysteine researchers agree that cholesterol was grabbing attention, but they doubt that there was any organized attempt to squelch investigations in other areas, because NIH was still funding homocysteine studies. Plus, says Finkelstein, excess homocysteine might only account for a small percentage of cardiovascular disease cases. "Homocysteine was too small to be a problem. The homocysteine people were not even on the screen for the fatheads," he adds, referring to cholesterol researchers.

Undeterred, McCully reiterated his ideas in a 1975 paper in the journal *Atherosclerosis.* The report described feeding experiments with rabbits, as had been done to study cholesterol, but it was the paper's title that had a lasting impact: "The Homocysteine Theory of Arteriosclerosis." By capturing an entire hypothesis in a title, McCully was in effect staking claim to the idea, because he had, in fact, made the initial conceptual link. But when his NIH grant was not renewed, he ran into trouble keeping his laboratory space and appointment at Harvard. He spent two years unemployed, until finding in 1981 the position that he still has today as chief of pathology and laboratory medicine at the Veterans Administration Medical Center in Providence, Rhode Island. Whether McCully's career difficulties were due to his NIH grant woes, the intense popularity of the cholesterol theory, or the fact that he wasn't trained to answer the epidemiologic questions his work had raised remains a mystery and a topic other homocysteine researchers like to avoid. But McCully's work would resurface.

Demonstrating that elevated homocysteine level or deficient vitamins could cause cardiovascular disease required large-scale clinical trials, hardly the turf of a pathologist with training in biochemistry, molecular biology, and genetics. Another impediment to providing evidence to sustain the homocysteine theory of arteriosclerosis was the crude analytical technology then available to monitor lower levels of plasma homocysteine. But at the peak of McCully's difficulties with Harvard, a key investigation proceeding on the other side of the world would support his extrapolation and open the way for the many studies that would eventually catalyze the transition of the homocysteine hypothesis to a supported theory.

A FLOOD OF STUDIES LINKS HIGHER HOMOCYSTEINE LEVEL TO CARDIOVASCULAR DISEASE

David and Bridget Wilcken, a cardiologist and pediatrician at Prince of Wales Hospital in Sydney, Australia, applied McCully's idea that high homocysteine concentration raises the risk of heart disease to people under age 40 who suffered from coronary artery disease, comparing them with healthy matched controls. All participants were given a large oral dose of methionine to see if this would stimulate buildup of homocysteine. Following this "methionine challenge," homocysteine levels in blood plasma were significantly higher among the people with coronary artery disease, suggesting that they could not metabolize the amino acid as efficiently as those with healthy hearts. This observation was important because it effectively extended the role of elevated homocysteine level beyond the extremely rare inborn errors of metabolism to more common forms of cardiovascular disease. Perhaps a less extreme elevation in homocysteine level may explain why otherwise normal people develop vascular disease.

The Wilckens' 1976 paper stimulated interest in homocysteine, but the pace of research was still hampered by the inability to detect low blood and urine levels of the very reactive amino acid. "Progress in studying this disease has reflected the state of technology as well as intellectual leaps," Finkelstein says. Donald Jacobsen, director of the Laboratory for Homocysteine Research at the Cleveland Clinic Foundation, agrees. "The assay for homocysteine from total plasma wasn't possible until 1985. It's a combination of radioimmunoassay and high performance liquid chromatography, and is very complicated. There are four different forms of homocysteine in serum. To get them all, hydrogens must be added to bond to and stabilize the sulfur groups so that pure homocysteine can be separated and measured," he says. The amino acid is very reactive and exists in plasma in several forms, called mixed disulfides; the free homocysteine form is only present for about two hours after eating. Jacobsen and his coworkers developed one of the modern versions of the assay by 1989. "Then, the papers came at a fast and furious pace," he adds.

There were studies. And then studies of studies, the meta-analyses. Then review articles summarizing the studies of studies. The dozens of investigations probed all sorts of individuals: young men with strokes, postmenopausal women with heart disease, children with family histories of coronary artery disease, healthy elderly people, and men and women undergoing coronary angiography (Table 8.4). A boost to the long-awaited analysis was that several ongoing large-scale clinical investigations of cardiovascular health were able to add homocysteine measurements to their protocols. And as the 1990s progressed, the data poured in.

But no matter how fast and furious the pace of homocysteine research, the perceived competitor cholesterol was still way ahead. In 1984, the NIH released its "Consensus Development Conference Statement on Lowering Blood Cholesterol to Prevent Heart Disease." Three years later, the AHA began the National Cholesterol Education Program, which publicized the results of the NHLBI study. Not surprisingly, the cholesterol-lowering drug market exploded.

Overall, study results supported the correlation between elevated homocysteine level and increased risk of heart and blood vessel disease. But it was slow going. "Only after there were prospective studies linking blood levels of homocysteine with future risk of myocardial infarction [heart attack] did the field really accelerate. The early evidence for an association was really not convincing. These things take a lot of time—for cholesterol, it took a few decades," says Meir Stampfer, a professor of epidemiology and nutrition at the Harvard School of Public Health.

Stampfer headed the first prospective clinical trial to track homocysteine level over time, the Physicians Health Study, published in 1992. An initial homocysteine level was taken for 14,916 male physicians who did not have atherosclerosis, and then they were reevaluated 5 years later. A plasma homocysteine level in the top 5%—equaling 12 micromoles per liter (μmol/L) or greater—correlated to a threefold increased risk of heart attack. "It was pretty easy to use the Physicians Health Study. We had already collected the blood, and just needed a good lab for the assays," recalls Stampfer.

In Norway, the Tromso study on 21,826 men and women found the relative risk of coronary artery disease to increase by 1.32 (about a one-third increase in risk) for every 4 μmol per liter of plasma homocysteine. And the famed Framingham Heart Study, begun in 1949 to follow the health of members of a Massachusetts community, looked at 1041 patients over the age of 67 and determined that elevated homocysteine level was an apparent risk factor for heart disease independent of the other risk factors. All the studies revealed a graded response; that is, there is no threshold level of homocysteine in the blood above which damage occurs.

Another epidemiologic approach was to compare homocysteine levels between healthy individuals and those with certain types of cardiovascular disease. The association persisted. The British Regional Heart Study, for example, found that of 5661 middle-aged men followed for 13 years, the 109 who suffered strokes tended to have higher homocysteine levels. Various meta-analyses concluded that about 10% of coronary artery disease in the general population could be associated with elevated homocysteine level. And as the data accrued, the homocysteine theory of arteriosclerosis evolved from "an obscure

TABLE 8.4

Clinical Trials That Established a Link Between Homocysteine Level and Cardiovascular Disease Risk

Study	Participants	Outcome
1999		
Framingham Heart Study	1933 healthy elderly people	Elevated homocysteine level correlates to increased overall death rate and increased death rate from CVD
Women's Health Study	122 women with CVD; 244 matched controls	2.6-fold increase in death rate due to CVD
University of Maryland	167 women with stroke; 328 matched controls	Women in top 40% of homocysteine level had 2.3-fold increase in risk of stroke
Bogalusa Heart Study	154 children whose parents have CAD; 983 matched controls	Level of homocysteine was increased in children with family history of CAD
1998		
Leyenburg Hospital, The Hague, and University Hospital, The Netherlands	89 people with recurrent blockages in veins; 227 matched controls	High doses of vitamins lowered homocysteine level in both groups
European COMAC Group	750 people with vascular disease; 800 matched controls	>80th percentile of plasma homocysteine or deficiency of folic acid or vitamin B_{12} was associated with increased risk of vascular disease
1997		
Erasmus University Medical School, Norway	131 people with >90% blockage in a coronary artery and >40% blockage of a second artery; 88 people with <50% blockage in only one artery; 101 healthy matched controls	Positive correlation between level of homocysteine and degree of blockage, with no threshold level
University of Washington School of Public Health	79 young women with MI; 386 matched controls	Those in the 90th percentile of plasma homocysteine have 2.3-fold increased risk of MI; high homocysteine level linked to low folic acid
1996		
European Union Concerted Action Project	750 people with vascular disease; 800 matched controls	Plasma homocysteine >12 µmol/L doubles risk of MI, stroke, vascular disease

CAD = coronary artery disease; CVD = cardiovascular disease; MI = myocardial infarction.

hypothesis to a major current topic in preventive cardiology," wrote Stampfer in a 1997 editorial in *American Family Physician.*

THE VITAMIN CONNECTION: FOCUS ON FOLATE

The various studies of homocysteine fell into two categories: the aforementioned ones that detected elevated homocysteine level among people who develop cardiovascular disease, and studies that linked increased vitamin intake to lower homocysteine levels. History seemed to be repeating itself, with homocysteine following the same steps that had been taken to implicate excess cholesterol in damaging blood vessels.

In the 1960s, the NIH sponsored the Diet-Heart Feasibility Study, in which 1000 men aged 45 to 54 followed either of three diets: two low in fat, the third consisting of more typical fatty American fare. Serum cholesterol fell 11% to 12% in the low-fat groups, but just 3% in the other. (That it fell at all in the third group attests to how awful the average American diet probably is!) Clearly, lowering dietary fat could lower serum cholesterol, and the Lipid Research Clinics Coronary Primary Prevention Trial would, in the next decade, further correlate this action to improved health. Similarly, by the mid-1990s, it was widely known that dietary intervention could lower levels of homocysteine too, but the outcome on health was not yet demonstrated. Research needed to go one step farther and examine the effects of increasing vitamin intake on both homocysteine level and health.

Dietary studies of homocysteine focused on folic acid and vitamins B_6 and B_{12}. Recall that folic acid supplies methyl groups, and B_6 and B_{12} are cofactors for key enzymes in the homocysteine pathway. The Homocysteine Lowering Trialists' Collaboration meta-analysis from Radcliffe Infirmary in Oxford, United Kingdom, for example, analyzed 12 studies and concluded that folic acid lowers homocysteine by 25%, B_{12} lowers it by 7%, and B_6 does not lower it at all. Much attention has since focused on folic acid, a vitamin important to health for several reasons.

Folic acid is a vitamin for which the statement "a synthetic and a natural vitamin are the same thing" does not apply. The natural form, folate, is a mixture of several compounds with the same overall chemical activity as the synthetic form, pteroylmonoglutamic acid. The practical difference, though, is that synthetic folic acid has twice the bioavailability as the folates found in foods and is much more heat stable, meaning that it gets into the bloodstream much more easily. Therefore, synthetic folic acid added to foods such as grains is likely to have a far greater effect on a person's health than folate obtained from foods. More important, the terms *folic* acid and *folate* cannot be used interchangeably, for the former is twice as potent as the latter. This has led to much confusion, even among physicians. For example, 200 micrograms (μg) of folic acid is required to correct a form of anemia, but the same therapeutic action would take 400 μg of folate obtained from food.

The Food and Drug Administration (FDA) issued a regulation in 1996 requiring that 140 μg of folic acid be added to every 100 g of grain products. The ruling went into effect in January 1998, with folic acid added to enriched flour, rice, pasta, and cornmeal. But the reason for this fortification had nothing to do with homocysteine. For many years, various studies had noted a protective effect of folic acid against neural tube defects, such as spina bifida and anencephaly. The United States began adding folic acid to grain products to prevent these birth defects, but the action has also lowered blood homocysteine levels in many people. This unanticipated benefit showed up in studies published in 1998 and 1999.

Irwin Rosenberg and his colleagues at Tufts University examined blood samples from the Framingham Offspring Study taken before and after fortification began, and discovered that homocysteine levels dropped as folic acid levels rose. In another trial, René Malinow and coworkers at the Oregon Health Sciences University gave patients with coronary artery disease Total breakfast cereal with three different levels of folic acid. The study was double blinded, so that neither participants nor researchers knew who was eating which cereal. Plus, each participant tried all three cereals at different times. The results were startlingly clear: As plasma folic acid increased, plasma homocysteine decreased. Eating one cup of regular Total cereal a day for at least 15 weeks, the study found, could reduce the blood homocysteine level by 11%. The relationship was proportional—the higher the folic acid, the lower the homocysteine.

Despite the results of the breakfast cereal study, products cannot yet tout the health benefits of consuming folic acid to lower homocysteine until the ongoing prospective clinical trials are completed and the results indicate that this is in fact beneficial. When a natural foods company petitioned the FDA in 1997 to approve a label stating "B-complex vitamins—folic acid, vitamin B_6, vitamins B_{12}—may reduce the risk in adults of cardiovascular disease by lowering elevated serum homocysteine levels, one of the many factors implicated in that disease," the agency disallowed the claim, even though by the time of the FDA's ruling in June 1998, the clinical trials that may finally demonstrate the link to improving health were already under way.

ESTABLISHING CAUSALITY AND MECHANISM

For people with a personal or family history of cardiovascular disease, the association between homocysteine and disease is already clear. In light of this, a year after folic acid fortification began, the AHA issued a science advisory urging physicians to test such patients for elevated blood homocysteine level and to advise all patients to eat more foods that contain folate and vitamins B_6 and B_{12}. Said Malinow, lead author of the advisory, "It will probably soon be as common to have one's homocysteine level checked as it is now to have one's cholesterol level checked. Folic acid, vitamins B_6 and B_{12} are likely to become significant weapons in the war against heart disease."

The reason that many researchers, physicians, and the AHA do not advocate routine screening of *everyone* for homocysteine is that causality has not yet been established. "Studies to date have not proved conclusively that there is a predictive relationship between the level of homocysteine in the blood and the likelihood of heart attack or stroke in adults who are otherwise at low risk for these diseases. Even if we could establish a predictive role for homocysteine, we have no data to show that lowering homocysteine will reduce an individual's risk of developing cardiovascular disease at a later date," says Ronald M. Krauss, a member of the AHA's nutrition committee and head of the department of molecular medicine at the University of California at Berkeley.

But demonstrating causality may only be a matter of time. Says Graeme J. Hankey, a neurologist at Royal Perth Hospital in western Australia, "I think the reason that it has taken so long for the homocysteine-heart health link to be validated and accepted is because it has taken time to conduct methodologically rigorous studies and to repeat them and find reasonably consistent results. It is certainly not proven that homocysteine actually causes symptomatic atherosclerotic vascular disease. It is still possible that the

TABLE 8.5

Demonstrating Causality
1. Do differently designed studies yield the same conclusions?
2. Is the relationship between the risk factor and the outcome maintained over time?
3. Is the risk factor independent, that is, not influenced by other risk factors?
4. Does altering the risk factor alter the risk?
5. Does a logical biological mechanism explain the observed relationship between the risk factor and the health outcome?

link is coincidental or a chicken-and-egg effect—that is, that high homocysteine is caused by the vascular disease."

Kilmer McCully, however, believes that epidemiology has already shown causality. He points to a 100-year perspective on death rates from heart disease released in 1999 from the Centers for Disease Control and Prevention (CDC) that indicates a drop from 307.4 cases per 100,000 population in 1950 to 134.6 cases by 1996. The report attributed the decline to controlling hypertension, changing the fat content of the diet, less smoking, and more exercise. "A lot of these factors are beneficial, but they didn't start until the mid to late 1970s," points out McCully, and are therefore inadequate to account for fewer heart disease deaths. Instead, he notes the increase in vitamin intake that has paralleled the drop in heart disease incidence. "The information is already out there that proves the validity of the theory. Adding vitamins to the food supply is why the death rate from coronary heart disease dropped precipitously in the mid 1960s. There is no other explanation. A meeting in 1979 examined this trend, with experts from all over considering smoking, exercise, diet, surgical intervention, and concluded that they couldn't explain why heart disease incidence was dropping. The reason was that first vitamin B_6, and later folic acid, began to be added to the food supply in voluntarily fortified cereals. That's when the decrease in heart disease became evident," he adds.

But folic acid, the most important of the three vitamins in maintaining cardiovascular health, was not added to grains in appreciable amounts until 1998, two years after the latest CDC data became available. More philosophically, the simultaneous occurrence of vitamin fortification and falling heart disease rates is again a correlation, not a cause. Even against the backdrop of the dozens of studies showing that people with cardiovascular disease are more likely to have elevated blood homocysteine level than others, large-scale clinical trials are needed to address the causality question (Table 8.5). They are ongoing in the United States, Norway, the United Kingdom, Australia, Spain, Finland, Ireland, the Netherlands, and elsewhere.

Within three to five years, results from the trials are expected to fill in the blanks, revealing whether lowering homocysteine level by consuming more folic acid reduces the risk of disease, just as the Lipid Research Clinics Coronary Primary Prevention Trial demonstrated for cholesterol. Explains Jacobsen, "It has been difficult for the clinical and cholesterol people to accept homocysteine as a major risk factor. But, the way that we practice medicine is evidence-based. We have to demonstrate that homocysteine is causal before it will be accepted as a major player. We await the results of the clinical trials that could show that. It is a truly international effort of at least 12 trials and 70,000 participants combined."

Some of the studies are primary intervention, considering healthy individuals, and some are secondary intervention, following people who have had a heart attack, stroke, or peripheral vascular disease to see if lowering homocysteine lowers recurrence risk, suggesting protection. The most meaningful data may come from the European and Australian trials, where folic acid is not added to grains, and vitamin supplements must be obtained by prescription. In the United States, establishing appropriate control groups would mean depriving people of a micronutrient that is otherwise part of the diet, something an Institutional Review Board would probably not approve.

If the ongoing clinical trials show that people with lowered homocysteine level live longer, testing for it will indeed become as common as cholesterol testing. Many in the field expect this to happen. "Too many recent studies show homocysteine level is important to survival, particularly if you already have coronary artery disease. Getting adequate folic acid and vitamins B_6 and B_{12} almost guarantees lowering homocysteine level by 10 to 40 percent. But we have to get the evidence. We're all waiting to find out if we're still in business," says Jacobsen.

In a practical sense, the probable outcome of the trials will likely be to extend the recommendations to take folic acid and perhaps B_6 and B_{12} to most of the general public. It's an easy, safe, and inexpensive treatment, albeit one that will not make fortunes for pharmaceutical companies the way that such cholesterol-lowering drugs as Mevacor, Pravachol, Zocor, and Lipitor have. An alternative finding, though, might be that elevated homocysteine is instead a marker, a biochemical bystander of sorts to whatever agent actually damages arteries. So although the end of the homocysteine story for most people may simply be to get more vitamins or eat more vegetables, for biologists, the closing chapter will be the revelation of *how* excess homocysteine does what it does—the Holy Grail of research, uncovering the mechanism.

So far, we don't have enough information to point to one specific way that homocysteine damages arteries. Laurence Pilgeram's long-ago hypothesis directly implicated methionine and a disruption in methyl transfer that roughens blood vessel linings, perhaps set into motion by homocysteine accumulation. But homocysteine may have a more direct role too. Current thinking is that homocysteine oxidized in blood plasma releases reactive oxygen compounds (free radicals) that tear up the endothelium, exposing the smooth muscle beneath it (Figure 8.6). The irritated muscle cells divide, creating bumps that snare passing lipids, while also activating platelets that make the blood more likely to lay down fibrin strands and to clot. The presence of homocysteine and its various sulfur-containing relatives oxidizes LDL particles, making them more likely to aggregate and become the first fatty deposits on the artery walls.

One candidate for such a damaging metabolite is homocysteine thiolactone, a short-lived derivative seen only *in vitro* that gobbles up the amino (NH_3) groups of other amino acids. "Any conditions that elevate homocysteine, or intracellular homocysteine/methionine ratios, also lead to an increase in thiolactone levels and in the degree of protein damage," suggests Hieronim Jakubowski, an associate professor in the department of microbiology and molecular genetics at the University of Medicine and Dentistry of New Jersey. The role of thiolactone in damaging blood vessels, however, has yet to be shown *in vivo.*

Finkelstein is placing bets on adenosylhomocysteine as the true risk factor. "This is an intermediate in the conversion of methionine to homocysteine and it has the

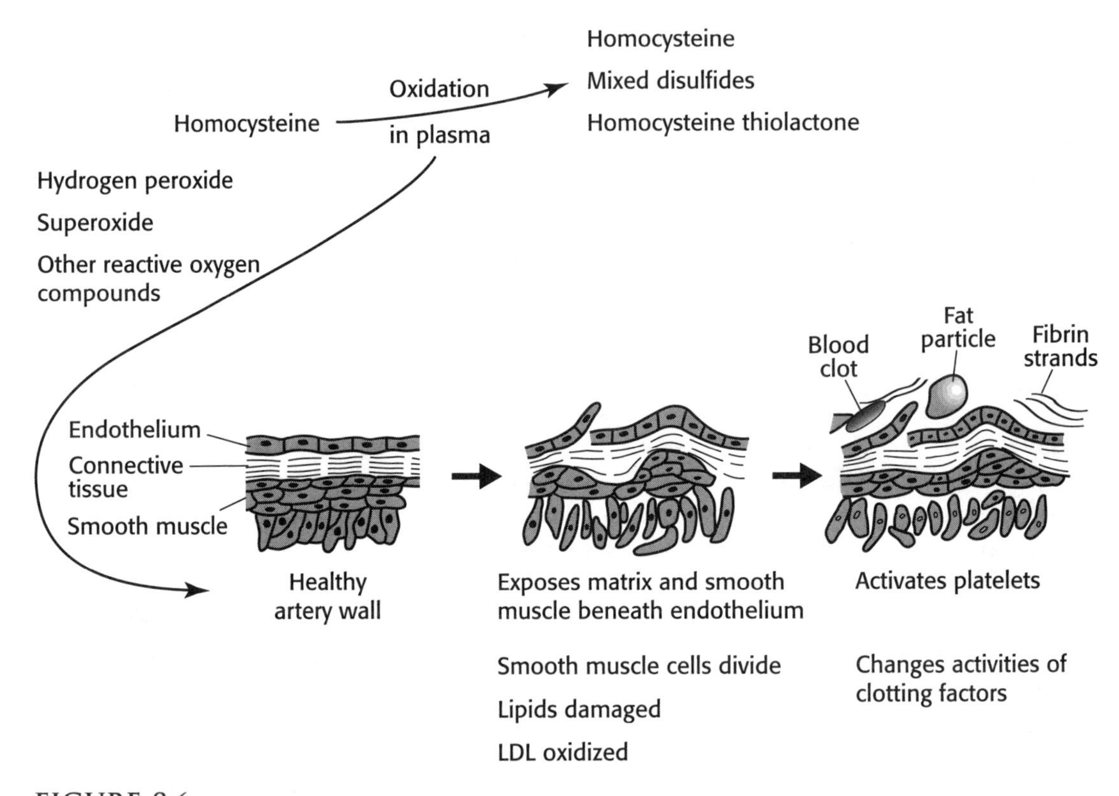

FIGURE 8.6

One explanation for how high homocysteine levels damage the cardiovascular system is the production of oxygen free radicals, which set into motion a chain reaction of tissue destruction.

important property of being elevated in all forms of homocystinuria," he explains. Adenosylhomocysteine can block activated methionine from releasing its methyl group. "I am backing this horse as the toxic intermediate, but I would lose little sleep if it turned out to be thiolactone," he adds.

Like reconstructing the details of a crime scene from pieces of evidence, researchers can assemble the probable steps in homocysteine's assault on arteries, but whether the scenario is accurate isn't known. Perhaps it is the combination of an elevated homocysteine level and eating a fatty diet that sets the stage for heart disease.

THE BIGGER PICTURE

The story of homocysteine research illustrates several themes of discovery in biomedical science. As in the stories of telomeres, stem cells, prions, and cloning, the research process is slow, incremental, and painstaking, accomplished by many and sometimes credited to few. "The homocysteine story has arrived at its present state due to the cumulative contributions of many workers who, each based on their own particular backgrounds and areas of scientific knowledge and interest, added to the synthesis," says Harvey Mudd of the laboratory of molecular biology at the National Institute of Mental Health.

Also like the other tales of discovery, that of homocysteine has its personalities: researchers bitter at having been ignored, and others who feel that the media have repeatedly painted only a partial portrait. Perhaps because science and celebrity do not mix well, Kilmer McCully's visibility has spawned tension within the homocysteine research community, although few will freely comment on it.

The popularization of the homocysteine theory of arteriosclerosis is a fairly recent phenomenon and not yet very well known, but much of what the public does know can be credited to McCully's cooperation with the media. The relationship began when several of the large-scale clinical trial results published in the mid-1990s cited McCully's 1969 paper in the very first sentence. By 1995, with abundant data now connecting elevated homocysteine level and increased risk of cardiovascular disease, McCully was introduced as the "father of homocysteine" at the first International Conference on Homocysteine Metabolism held in County Clare, Ireland.

The New York Times couldn't resist the idea of one person founding a field, and so their July 4 coverage of the conference included comments from several researchers but focused on McCully, whose photograph accompanied the article. By the time that *The New York Times Magazine* published "The Fall and Rise of Kilmer McCully" on August 10, 1997, the exclusive link between the man and the molecule was effectively sealed. Wrote author Michelle Stacey, "Homocysteine is the theory of the moment in a notoriously competitive field, and McCully's name is inextricably tied to its ascent." Many scientists would argue that a field dating from the 1930s is not quite as recent or ephemeral as that comment suggests; however, the fact that McCully had become its unofficial guru was not in much doubt. In 1997, he had taken his story to the masses himself by writing a book, *The Homocysteine Revolution.*

But it wasn't only the mainstream press that celebrated McCully. In 1998, a news feature in the *Lancet* profiled "Kilmer McCully: Pioneer of the Homocysteine Theory." This article continued the juxtapositioning of homocysteine and cholesterol as if they were locked in mortal battle for the honor of causing heart disease.

In 1999 came *The Heart Revolution,* McCully's second book, coauthored with daughter Martha. In the introduction, Michelle Stacey compares McCully to Galileo. The hyperbole isn't really necessary, for both books are well written, recount the history of homocysteine research, and are packed with the nutritional advice that the public craves. Both books also clearly credit those who worked out the pathway, attributions that seem to have been swept aside in the media quest for a superstar. The books sold, and soon McCully hit the talk show circuit, an activity that many scientists regard as self-promotional.

The image of one man as the lone battler against the evil homocysteine persists in magazine articles, some of which wander far from the reality of the homocysteine story in particular and the way that science is done in general. Consider an article in the December 1999 issue of *Discover* magazine entitled "What If Cholesterol Isn't the Basis of Heart Disease After All?" The piece describes the "rogue theory of heart disease progression, first proposed 30 years ago, [that] holds that another player is at work," completely missing the point of what an independent risk factor is. Less forgivable is that author Karen Wright wrote, "McCully discovered that the homocysteine molecule is metabolized by enzymes that act in concert with B vitamins," thereby dismissing and miscrediting the elegant work of many others, done over decades, in under a sentence!

The *Discover* article caused many members of the homocysteine research community to collectively cringe. Hieronim Jakubowski wrote to the magazine to correct this error, ending his unpublished letter with "Scientists are, by nature, very skeptical, but are easily convinced by solid data. Such data regarding homocysteine and vascular disease were largely missing until the 1990s. It was the efforts of hundreds of scientists that led to the acceptance of homocysteine's role in vascular disease."

Of course the important lesson to be learned from homocysteine research will be an enormous public health benefit, if the ongoing clinical trials indicate that lowering levels of this amino acid can indeed prevent heart disease. And it is a wonderful story, because prevention is so simple. But the larger story is the recurring one of scientific discovery, the fact that despite the media's portrayal of science as a series of breakthroughs perpetuated by a few individuals, learning how nature works is almost always the fruit of many interacting minds.

REFERENCES

Boushey, C. J., et al. "A quantitative assessment of plasma homocysteine as a risk factor for vascular disease." *Journal of the American Medical Association* 274:1049–1057 (1995).

Brody, Jane. "Health sleuths assess homocysteine as culprit." *New York Times*, 13 June 2000, F1.

Clarke, R., and R. Collins. "Can dietary supplements with folic acid or vitamin B_6 reduce cardiovascular risk? Design of clinical trials to test the homocysteine hypothesis of vascular disease." *Journal of Cardiovascular Risk* 5:249–255 (1998).

Garrod, A. "The lessons of rare maladies." *The Lancet* 1:1055 (1928).

Graham, Ian, and Raymond Meleady. "Heart attacks and homocysteine." *British Medical Journal* 313:1419–1420 (1996).

Hankey, Graeme J. "Homocysteine and vascular disease." *The Lancet* 354:407–413 (1999).

Jacques, Paul F., et al. "The effect of folic acid fortification on plasma folate and total homocysteine concentrations." *New England Journal of Medicine* 340:1449–1454 (1999).

Lewis, Ricki. "Homing in on homocysteine." *The Scientist*, 24 January 2000, 1.

Malinow, M. R., et al. "The effects of folic acid supplementation on plasma total homocysteine are modulated by multivitamin use and methylenetetrahydrofolate reductase genotypes." *Arteriosclerosi, Thrombosis and Vascular Biology* 17:1157–1162 (1997).

McCully, K. "Vascular pathology of homocysteinemia: Implications for the pathogenesis of arteriosclerosis." *American Journal of Pathology* 56:111–121 (1969).

McCully, K., and R. B. Wilson. "The homocysteine theory of arteriosclerosis." *Atherosclerosis* 22:215–227 (1975).

McCully, Kilmer. *The Homocysteine Revolution.* New Canaan, CT: Keats Publishing, 1997.

McCully, Kilmer, and Martha McCully. *The Heart Revolution.* New York: HarperCollins, 1999.

Meleady, Raymond, and Ian Graham. "Plasma homocysteine as a cardiovascular risk factor: Causal, consequential, or of no consequence?" *Nutrition Reviews* 57:299–305 (1999).

Mudd, S. H., J. D. Finkelstein, F. Irreverre, and L. Laster. "Homocystinuria: An enzymatic defect." *Science* 143:1443–1445 (1964).

Pilgeram, L. "Susceptibility to experimental atherosclerosis and the methylation of ethanolamine 1,2-C14 to phosphatidyl choline." *Science* 120:760–765 (1954).

Stampfer, M. J., et al. "A prospective study of plasma homocyst[e]ine and risk of myocardial infarction in US physicians." *Journal of the American Medical Association* 274:1049–1057 (1992).

Wilcken, D., and B. Wilcken. "The pathogenesis of coronary artery disease: A possible role for methionine metabolism." *Journal of Clinical Investigation* 57:1079–1082 (1976).

ON TECHNOLOGY: GENE THERAPY AND GENETICALLY MODIFIED FOODS

Most people become aware of science, other than in the classroom, only when it directly affects their lives. This might be an event, such as a devastating storm or illness, or an application of a new technology. Would anyone other than a biologist, for example, notice that a fleck of fungus fallen on a plate of bacteria kills the bacteria? Yet the same fungal toxin, as the antibiotic drug penicillin, became known to all when it saved millions of lives during World War II, and it and its pharmaceutical descendants continue to be part of our arsenal against infectious disease.

The hot topics chronicled in this book have been compared to icebergs: The bulk of the stories are hidden, with only the exposed parts attracting media attention. Few people had heard of telomeres, for example, until newscasters likened them to a fountain of youth. Biologists recognized the potential of stem cells to yield replacement parts for decades, but the cells didn't enter the public lexicon until two key papers propelled politicians and ethicists into arguing over the implications. Homocysteine research has plodded along in the shadow of the almighty cholesterol, with the concluding chapter very likely to be the decidedly low-tech admonition to "eat your vegetables." Another example of a field of life science that was largely ignored, with the exception of science fiction treatments, until it was very far along is cloning. To most people, cloning appeared suddenly as a full-fledged and frightening technology, not as the predictable application of developmental biology that it is.

When technology evolves from a life science, society seems to respond with either great expectation or fear. On the one hand, we have come to expect the utter contradiction of daily, or perhaps weekly, breakthroughs. Yet at the same time, people have both a fear of the technical and a fear of the unknown, setting the stage for rejection, or even panic, in the face of new ways of doing things. Two biotechnologies in particular, each with roots in the 1980s but in the news now, illustrate these opposing reactions of unrealistic expectations and perhaps unwarranted fears: gene therapy and genetically modified organisms used as food.

THE FIRST REPORTED GENE THERAPY DEATH

No one expected Jesse Gelsinger to die. The 18-year-old man succumbed, in a matter of days, in September 1999, from an overwhelming reaction of his immune system to an experimental gene therapy. Jesse's death was especially disturbing because he seemed to have had his inborn error of metabolism—ornithine transcarbamylase (OTC) deficiency—under control, even though that meant careful attention to diet and taking nearly three dozen pills a day. In addition, his case was milder than most because he was a mosaic. That is, only some of his cells couldn't manufacture the enzyme in question, OTC (Figure 9.1).

OTC is one of five enzymes that are part of a metabolic pathway called the urea cycle. The enzyme is necessary to break down the ammonia (NH_3) groups that are liberated as the body digests dietary protein into amino acids. Without the enzyme, ammonia builds up rapidly in the bloodstream, and when it accumulates in the brain, coma and death result. Most affected individuals are newborns, who become comatose within 72 hours of birth. Half of them die by 1 month old, and 25% more by age 5 years. Those who survive can control their symptoms by taking drugs that bind ammonia and by following a special low-protein diet—as Jesse did.

OTC deficiency is X linked, which means that carrier mothers pass it to each of their sons with a probability of 50%. One in 40,000 newborn boys is affected. It was for these children, these babies, that Jesse Gelsinger wanted to volunteer for a gene therapy trial to be held at the University of Pennsylvania. He knew that he would not personally benefit, saying shortly before his first and fatal treatment, "What's the worst that can happen to me? I die, and it's for the babies." His father, Paul, had heard of the planned experiment shortly after Jesse went into a coma in December 1998 from failing to take all his pills. When Jesse recovered, his father told him about the gene therapy, and they contacted the head of the clinical trial, Jim Wilson, director of the Institute for Gene Therapy at the University of Pennsylvania. Wilson insisted that Jesse wait until he was 18, to be sure that he fully understood the possible risks of this new approach to treating disease.

The gene therapy consisted of an adenovirus—a type of virus that causes the common cold—that harbored a working human OTC gene. Mark Batshaw, chief academic officer at Children's National Medical Center in Washington, had designed the therapy and chose this virus to deliver the healing gene because it is large and travels quickly to the liver, where it is needed in OTC deficiency. A fast fix would be necessary to help newborns within 72 hours. Also, disabled adenovirus had already been used in about a quarter of the 330 gene therapy experiments done on more

FIGURE 9.1

Jesse Gelsinger died at age 18 from gene therapy to treat an inborn error of metabolism, ornithine transcarbamylase deficiency.

Photograph courtesy of the Arizona Daily Star. *Used with permission from the Gelsinger family.*

than 4000 patients since 1989. The only apparent problem was an occasional transient immune response, a reaction that had sidelined a cystic fibrosis gene therapy trial some years ago. The adenovirus used in the OTC deficiency trial, as in many others, was a stripped-down version that had the gene-transferring capability maintained but the genes that enable it to replicate and cause disease removed.

The experimental protocol called for three groups of six patients to receive three different doses of the engineered virus. University of Pennsylvania bioethicist Arthur Caplan had urged that it not be tried on newborns because their parents would be too shocked and distraught to make informed decisions. Instead, the volunteers were young men with the disease, or women who were carriers. The clinical trial would, the researchers hoped, identify the maximum tolerated dose—that is, the amount of doctored virus that is high enough to deliver enough copies of the gene to ameliorate symptoms, yet low enough not to cause serious side effects, toxicity, or an immune response. Jesse was the eighteenth and last patient to be treated, and he was in the highest dose group. The other participants had done well. Two patients suffered a high fever and aches and pains, but these symptoms were not considered serious enough to halt the trial.

What happened to the young man was shockingly fast. Shortly after Jesse entered the hospital on Monday, September 13, 3.8 trillion engineered viruses were sent into an artery leading into his liver. When he developed a fever at night, no one was particularly alarmed because an initial immune response of a fever to a strange virus is not unusual. By morning, however, the clinical picture had changed. Jaundice had set in, the buildup of bilirubin released from shattered red blood cells already turning the whites of Jesse's eyes and his skin a sickly yellow. His liver could not keep up with breaking down the hemoglobin spewing from the burst red blood cells, and the overload of that protein sent the ammonia level in his liver and elsewhere soaring past 10 times the normal level by midafternoon. His blood also was not clotting normally. Jesse became disoriented, then comatose. By Wednesday, doctors had managed to control the ammonia buildup in his bloodstream, but then his lungs began to fail, so they placed Jesse on a ventilator. On Thursday, Hurricane Floyd hit the East Coast, preventing some of the researchers from reaching the hospital where, one by one, Jesse Gelsinger's vital organs were shutting down. By Friday he was brain dead. His father, devastated because it was he who had suggested that Jesse enroll in the experiment, finally turned off the ventilator, and Jesse died.

Wilson and Batshaw reported the death. The news didn't, at first, make too much of a flutter in the media. Then a public hearing was held in early December as part of a meeting of the Recombinant DNA Advisory Committee of the National Institutes of Health (NIH). There, families from the National Urea Cycle Disorders Foundation, as well as Jesse's father, voiced their support of the gene therapy, imploring officials to allow it to continue because it was their only hope for a cure.

At the hearing, the doctors and researchers tried to interpret clues from the autopsy findings known at that point in the investigation. In Jesse's liver, the engineered adenovirus had gone not only to the hepatocytes, where they were expected to go, but also to macrophages that alert the immune system to begin its assault with inflammation. In addition, the virus had spread, quickly, well beyond the target tissue, to the lymph nodes, spleen, bone marrow, and elsewhere. The autopsy revealed another surprise: Jesse's bone marrow lacked red blood cell precursors, indicating that he may have had an undetected second condition. It was also news that his bone marrow showed signs of previous parvovirus infection. This virus could have recombined with the adenovirus in a way that evoked the immune response. In addition, when Wilson examined the adenovirus in Jesse's bloodstream, he detected duplicated DNA sequences not seen in the original engineered virus, although they were not in an expressed part of a gene. This meant that the virus had changed, as genetic material tends to do.

Over the next few weeks, the trial that claimed the life of Jesse Gelsinger came under close scrutiny, and deficiencies in the administration of the treatment gradually came to light. At first, the only potential problem seemed to have been a change in protocol in 1996 when the NIH's Recombinant DNA Advisory Committee had been temporarily disbanded. The Food and Drug Administration (FDA) stepped in and changed the delivery from an intravenous route to directly applying the virus to the liver—an attempt at limiting toxicity that backfired. As the investigation continued, FDA and NIH investigators cited delays in reporting side effects and in filing eligibility forms, and inadequate documentation of informed consent. Wilson insists that Jesse knew what he was doing when he volunteered. Then the NIH examined other gene therapy trials and uncovered extreme underreporting of adverse side effects. However, this apparent oversight might reflect the fact that researchers considered many of the adverse effects to have been due to the underlying disease and not the gene therapy. But until the NIH can determine exactly what went wrong in Jesse Gelsinger's case, the agency is inspecting facilities sponsoring gene therapy trials and is requiring weekly reports of adverse effects.

Researchers are also taking a closer look at the suitability of adenovirus as a gene therapy vector. It might not be as benign as its usual effect on humans—upper respiratory infection—might suggest. At least one gene therapy trial using rodents found adenovirus to cause severe chronic inflammation not unlike the reaction that Jesse suffered. Related viruses are known to cause systemic illness, with severe inflammation, in dogs, birds, and horses. Arabian foals in particular experience liver inflammation if they have an inherited type of immune deficiency. Perhaps Jesse Gelsinger had an undetected inherited condition that set the stage for the overwhelming infection by and response to the adenovirus.

Other gene therapy trials utilizing adenovirus were halted in the aftermath of Jesse Gelsinger's death, while researchers paused to ask new questions and develop new ways to screen patients and predict consequences of gene therapy. In the future, using data from the human genome project, researchers should be able to utilize microarrays (DNA chips) to more thoroughly screen prospective participants for many inborn errors, predisposing factors, and immune system variations that might signal a tendency to overreact to certain viruses.

For the short term, the Recombinant DNA Advisory Committee has urged better monitoring of the "narrow window" of adenovirus dose that is therapeutic but not toxic based on reports that the virus can be safe at one dose but become very dangerous at just a slightly higher dose. The committee also recommended that gene therapy researchers learn to speak the same language—that is, to measure the viral dose in the same units and to seek the same type of evidence of efficacy or toxicity. With the influx of information from the human genome project, it is vital that gene therapy be standardized, because many new applications are certain to be developed.

Jesse Gelsinger's death was shocking and disturbing in the specific context of gene therapy, which had a relatively clean safety record. But it was not so unexpected in the broader context of the risks inherent in testing any new method of treating disease. Gene therapy efforts so far have been mostly disappointing, introducing genetic changes that may work at the cellular level, but fail to alleviate symptoms. For the most part, gene therapy has done no harm, and it has had a few noteworthy successes. In fact, the success of a few early efforts may have fueled a continuing optimism that has, until now, overshadowed the disappointments. The therapy *can* work.

The first recipient of gene therapy was 4-year-old Ashanti DaSilva. Shortly after noon on September 14, 1990—nearly nine years to the day before Jesse Gelsinger would receive his gene therapy—Ashanti watched her own T cells (a type of white blood cell), bolstered with a functioning copy of the gene that she lacked, drip into her arm at the NIH hospital in Bethesda. Her liver cells could not manufacture an enzyme, adenosine deaminase (ADA), whose absence caused buildup of an intermediate compound in a biochemical pathway that attacked her T cells, thereby crippling her immune system. For a child with ADA deficiency like Ashanti, life is a series of serious infections. Although by the time of the first gene therapy patients could receive the enzyme directly, or even be cured with bone marrow transplants, researchers recognized that a genetic approach could provide a longer-lasting treatment. And so Ashanti, and a few months later an 8-year-old girl, Cynthia Cutshall, had volunteered to receive her own doctored cells as an experimental gene therapy. It seemed safe, and it was. But as a precaution, each girl also received the enzyme treatment to prevent infections. Success was measured at the cellular level.

The treatment worked: In each girl, a small fraction of T cells acquired the healing gene. But because white blood cells do not live for very long, the fix was short-lived. W. French Anderson and his colleagues went back to the drawing board and tailored the therapy to less specialized cells that might persist in the circulation long enough to exert a noticeable effect. In the next stage of the evolution of this particular gene therapy, three fetuses were identified who had inherited ADA deficiency, and at birth, the newborns were treated with engineered white blood cells taken from their umbilical cords. These were stem cells, destined to have a much longer lifetime, and spawn more daughter cells, than the mature T cells that had been manipulated in Ashanti and Cynthia. The experiment seems to have worked. Over the years, the percentage of T cells that can produce ADA in the three children has steadily risen, and their infections have become more easily controlled.

Gene therapy successes have been few and far between, as the technology experiences growing pains. On June 1, 1999, Don Miller, a 50-year-old part-time math librarian and sheep farmer, became the first recipient of gene therapy for the blood-clotting disorder hemophilia A. He received a disabled virus that delivered a functional gene for clotting factor VIII to his bloodstream. It was a retrovirus, which has RNA as its genetic material and thus is different from the DNA-based adenovirus that was used to treat OTC deficiency in Jesse Gelsinger.

For Don Miller, living with a genetic disease hadn't been easy. His first symptom of hemophilia was at his circumcision, when he nearly bled to death. But doctors in his small town didn't know what he had until he was 18 months old—they'd never seen it before. Miller has very early memories of his illness. "When I was 3, I fell out of my crib and I was black and blue from my waist to the top of my head. The only treatment then was whole blood replacement. So I learned not to play sports. A minor sprain would take a week or two to heal. One time I fell at my grandmother's house and had a one-inch long cut on the back of my leg. It took 5 weeks to stop bleeding, just leaking real slowly. If I moved a little the wrong way, it would open and bleed again," he recalls.

The doctors tried to transfuse Miller as seldom as possible, for fear of transmitting hepatitis and of straining his kidneys. Miller remembers meeting other kids in the hospital with hemophilia who died from kidney failure as a result of too frequent transfusions. Then hemophilia treatments became more targeted as the technology for separating blood components improved. Instead of whole blood, Miller began receiving just plasma and certain antibodies, and then cryoprecipitate, which is the factor VIII clotting protein pooled from many donors. It was the use of cryoprecipitate that exposed many people with hemophilia to the human immunodeficiency virus (HIV) in the early 1980s. "I lucked out. I took so little cryoprecipitate that I wasn't exposed to very much," speculates Miller. Today he is one of the oldest people with hemophilia who is HIV negative. By the late 1980s, the threat of HIV infection to people with hemophilia had diminished, both because the blood supply was finally screened and because factor VIII became available courtesy of recombinant DNA technology and was therefore virus free.

Miller's age and good health got him admitted to the gene therapy trial at the University of Pittsburgh. The treatment consisted of infusions on three consecutive days with weekly monitoring, a protocol established in animal studies. So far Miller hasn't experienced any side effects, but he has noticed improvements in living with hemophilia. "In the eight weeks before the infusion, I used eight doses of recombinant factor VIII. In the fourteen weeks since then, I've used three. As long as I don't let myself feel stressed, I don't have spontaneous bleeding. Incidents that used to require treatment no longer do. I've had two nosebleeds that stopped within minutes without treatment, with only a trace of blood on the handkerchief, as opposed to hours of dripping," Miller says.

Gene therapy is a technology that seems to be in the throes of birth, inching forward with the successes, stalling when effects are transient or ineffectual, and slipping backward when damage is done—which is thankfully rare. But most medical technologies seem to go through a stage when many variables and details must be worked out and understood. Because gene therapy fundamentally makes sense—replacing faulty instructions—it seems likely that it will ultimately become a widespread technology that takes human genome information into the realm of the practical. Perhaps gene therapy will go the way of transplantation medicine, with a slow and rocky start followed by spectacular success as a basic technique finds many applications.

WILL GENE THERAPY FOLLOW IN THE FOOTSTEPS OF TRANSPLANTS?

Transplanting a heart first became a real possibility, rather than the stuff of science fiction, with the advent of the heart-lung bypass machine in 1953. The device's ability to temporarily take over the job of oxygenating the blood allowed surgeons the time to detach and reattach blood vessels, the nuts and bolts of a transplant procedure. In the 1960s, Richard Lower and Norman Shumway at Stanford University founded the field by transplanting dog hearts, which was both a success and a failure. The operations were a success because the dogs generally lived a few weeks, indicating that the mechanical attachments were correct. But then they died when their immune systems violently rejected the foreign organs. The dog experiments convinced Shumway and Lower that more research was required to better understand tissue rejection before heart transplants could be attempted on humans with any hope of success.

But the idea of a dramatic new life-saving technology was seductive, and for one brash young heart surgeon, irresistible. Christiaan Barnard used Shumway's techniques but ignored his warning not to proceed until more was known about the risks, performing

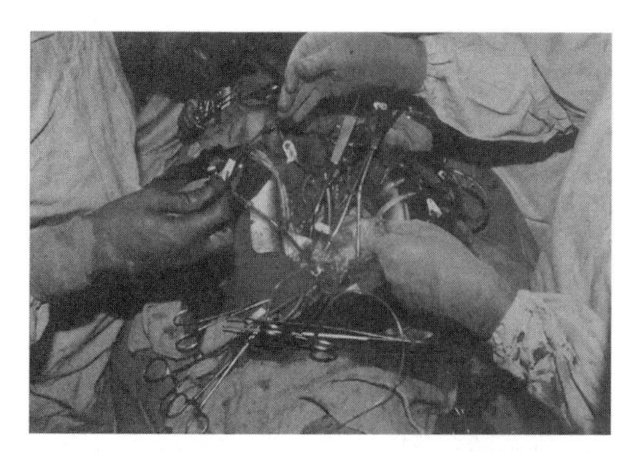

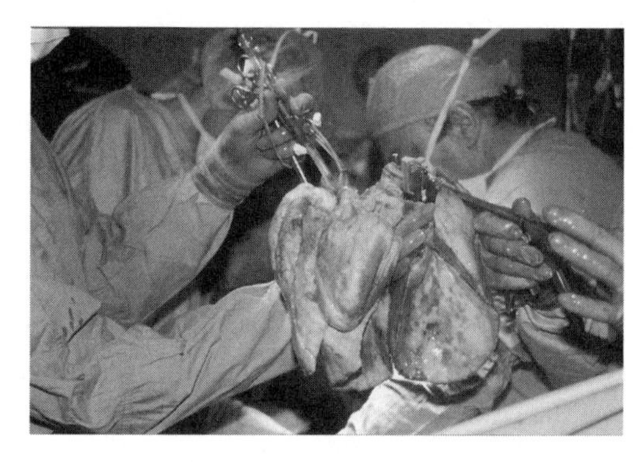

FIGURE 9.2

The first heart transplant was done at Groote Schuur hospital in Capetown, South Africa. The man lived 18 days. Today heart transplant recipients live for several years.
Courtesy of University of Wisconsin Organ Procurement.

the first heart transplant on December 3, 1967, at Groote Schuur Hospital in Capetown, South Africa (Figure 9.2). The operation on Louis Washkansky made worldwide headlines as it catapulted Barnard to fame. On December 20, Barnard took off for a whirlwind tour of the United States, where he was treated as a celebrity. Unfortunately, his patient died the next day, his body having violently rejected the organ. Meanwhile, Shumway and Lower's fears that the procedure was being done prematurely were realized when a surgeon in Brooklyn attempted a heart transplant on a 17-day-old infant, who lived only a few hours. Still, Barnard's transplant was widely regarded as a success, just as the initial dog experiments were, and he did a second operation on January 2, 1968. This patient, Philip Blaiberg, lived an astonishing 19 months with his new heart, and many recountings of heart transplant history tend to forget the unfortunate Mr. Washkansky, focusing on Blaiberg. But his success was a rarity in the early days of heart transplants.

A flurry of heart transplants was performed in 1968 (Table 9.1), but the excitement quickly diminished as survival times proved disappointing, usually measured in weeks. By 1971, only two of the world's 58 heart transplant teams were still in operation—Shumway's and Barnard's. Until the early 1980s, progress pretty much ground to a halt as researchers concentrated on understanding the immune rejection reaction. Then, a series of technological advances reawakened interest in heart transplants: the development of immunosuppressant drugs, improved surgical techniques, more precise ways to match donor organs to recipients, and the ability to strip molecules from the outsides of cells on donor organs, making them less likely to evoke an immune response. By 1984 in the United States alone, 29 centers transplanted more than 300 hearts.

Today's statistics on heart transplants are nothing short of amazing, with thousands of procedures performed annually and survivors living for years (Table 9.2). In the United States in 1997, the most recent year for which statistics are available, 2290 heart transplants were performed. About 40,000 individuals under age 60 awaited hearts. On New Year's weekend of 1995, Stanford University performed its thousandth heart transplant, to a 43-year-old man who also received a new pair of lungs. And in 1999, transplant recipients from 47 nations competed in the XII World Transplant Games, demonstrating vividly the success of transplantation medicine.

TABLE 9.1

Heart Transplants

Year	Number Worldwide
1968	99
1969	48
1970	17
1971	1

TABLE 9.2

Survival After Heart Transplant

Years	Percentage
1	82.5
2	78.2
3	74.4
4	70.5

Surgeons brave enough in the late 1960s and 1970s to attempt transplants probably could not have imagined how successful the approach would be today. Will the same be said for gene therapy 10 or 20 years from now? It's very likely.

Like transplantation medicine, gene therapy was an idea that had to await technology. Then, animal experiments showed the promise of the strategy for curing disease. Just as the first transplants failed or worked only for a short time because we did not fully understand tissue rejection, initial gene therapy failures also reflected an overzealous immune response. In both cases, the immune system is only doing what it does naturally—recognizing and attacking something not of the body. It is probably only a matter of time until we understand how that attack targets gene-carrying viruses and learn how to dampen the response.

A BUTTERFLY SYMBOLIZES TECHNOFEAR

Genes are genes, the genetic code of a plant the same as that of a bacterium, a mushroom, an ameba, or an aardvark. This universality of the genetic code makes possible the transfer of genes from one species to another. Recombinant DNA technology, in which single cells are given human genes and coaxed to secrete valuable proteins, stems from this idea. Although scientists and the public alike greeted this technology in the 1970s with trepidation, today a few dozen drugs, many unavailable any other way, owe their origins to bacteria, yeast, or mammalian cells genetically engineered to produce therapeutic proteins. Gene therapy is an outgrowth of this technology on a multicellular level.

Plant cells can, of course, be genetically manipulated too. Efforts to do so trace back about 10,000 years, in the form of the selective breeding that is traditional agriculture. Today, though, plants can be more precisely genetically modified with a gene of interest stitched into each cell, enabling the plant to produce a desired protein. An organism that harbors a gene from a different species is termed *transgenic.* A less precise but popular term is *genetically modified organism,* or GMO. Some GMOs have genes silenced, rather than added.

Modern agricultural biotechnology—altering, removing, replacing, or augmenting the genes of domesticated plants and animals—has existed since the 1980s, but only recently has it attracted much media attention. Reasons for current objections to genetically modified (GM) foods are complex, involving distrust of government regulators and scientists, politics, economics, lack of knowledge of biology in general and genetics in particular, and an apparent fear of combining genetic material from two types of organisms. The recent backlash against foods that contain GM crops originated in the United Kingdom in the late 1990s, with its roots in lingering suspicion from the days of mad cow disease earlier in the decade when the government misled the public concerning the safety of possibly prion-tainted beef (see Chapter 4). But to the U.S. biotechnology community, the European panic over GM foods was like déjà vu, what National Science Foundation president Rita Colwell calls a "been there, done that" feeling. We'd been through it before. Consider two similar events that occurred a dozen years apart.

In 1977, a graduate student at the University of Wisconsin, Steven Lindow, discovered that ice forms on the leaves of potato plants by clinging to proteins that are part of the surface of the bacterium *Pseudomonas syringae.* What would happen to the potato plants, Lindow wondered, if the bacteria could not produce the ice-nucleating proteins? Lindow first identified a naturally occurring bacterial mutant that made only a few of the proteins, then deleted the gene, which eliminated the proteins. Next, he tested these "ice-minus" bacteria on potato plants growing in a greenhouse. And he indeed found that leaves coated with normal bacteria froze at 28°F (–2.2°C), but that leaves coated with ice-minus didn't freeze until the mercury dipped to 23°F (–5°C). The mutant bacteria lowered the temperature at which ice could form on leaves.

Steven Lindow knew immediately that he had the microbial equivalent of a cash cow. A way to render crops frost-resistant could combat the average U.S. annual loss of $1.5 billion due to frost damage. The next step was to conduct a field test. It was 1982, and Lindow set his sights on Tulelake, California, near his job at the University of California at Berkeley. But a lawsuit from the Foundation on Economic Trends in Washington, DC, delayed the experiment for four years. The Foundation was and is the vehicle of Jeremy Rifkin, a nonscientist who has made a career of protesting and derailing biotechnology experiments. The objection to Lindow's proposed experiment was particularly odd because the genetic manipulation would *remove* a natural genetic function, not add a strange or foreign capability. While the Foundation painted images to the press of botanical monsters invading California, Lindow continued doing his science, testing the local plant life with ice-minus bacteria in greenhouses, studying local wind and pollen dispersal patterns, and talking to wildlife biologists as well as townspeople, who at first did not object to his planned experiment. Then in 1985, he hit another snag. A biotechnology company in Oakland, Advanced Genetic Sciences, tested similarly altered bacteria on trees growing on the rooftop of their facility, without obtaining a go-ahead from the Environmental Protection Agency, as Lindow had done. Although the company's "Frostban" experiment actually did no harm, it prompted scare headlines, and apprehension wafted north to Tulelake. The citizens began to question the safety of Lindow's experiment. In retrospect, the situation was eerily like today's debates—GM foods have harmed no one, but frighten many.

In the spring of 1987, after further research had confirmed the safety of the proteinless bacteria, Lindow finally planted his plants. He coated thousands of potato seeds with his ice-minus bacteria, then planted them and nurtured 3000 seedlings in a field in Tulelake. But early one morning, vandals ripped out or damaged half of the plants. The researchers replanted, this time spraying the seedlings with ice-minus bacteria in the field. The bacteria protected the plants from a late frost that year, and experiments showed that, even many months later, none of the 6 trillion released microbes had traveled beyond 32.7 yards (30 meters) on the naked soil surrounding the experimental plot. A version of the bacteria is still used, safely, today.

Lindow's experiment, a long time in coming, had ushered in the era of "deliberate release" experiments. And as was the case for recombinant DNA technology a decade earlier, precautions became less stringent as the years passed and the methods proved to be safe. When no one died from an attack of killer tomatoes, researchers abandoned the space suits they once routinely wore. The media ceased sounding alarms every time a researcher conducted a field test of a GM plant. But lack of protests does not make news, and somehow the absence of hysteria over deliberate release experiments of GM plants in the United States never made it to the United Kingdom.

After Lindow's experiment faded from the headlines, only a rare genetically modified plant or animal made news. Transgenic sheep that secrete a human clotting factor were nurtured at the same facility in Scotland that would house Dolly the cloned sheep years later (see Chapter 7). These sheep did not attract much media attention. One early casualty of agricultural biotech was the FlavrSavr tomato. Like the ice-minus bacteria, the FlavrSavr lacked a gene, one that encodes a ripening enzyme. The fruit ripens more slowly on the vine, extending its shelf life. The existence of the FlavrSavr would occasionally elicit howls of outrage when a consumer organization realized that the tomato was genetically altered yet allowed to sit in a supermarket bin masquerading as a normal tomato. But the FlavrSavr was doomed, both because of negative publicity and, its developers later admitted, a failure to pay attention to certain other characteristics of the plant that made it noncompetitive in the salad stakes. Old-fashioned home-grown Big Boy tomatoes and other varieties just tasted better. A hard lesson for agricultural biotechnology to learn was that new traits that ease the grower's life do not impress or even interest consumers. When it comes to food, people care about taste, and perhaps cost, not ease of processing.

By the late 1980s, the United Kingdom had its own version of the FlavrSavr, plus a beer made with genetically modified yeast. Neither got very far once consumers got wind of the "GM" part of the recipe. But the story of another product reveals how packaging means everything to a fickle and technophobic public.

In the early 1990s a British company introduced a cheese manufactured using a protein-digesting enzyme called chymosin. But their chymosin did not come from its natural source of calf stomachs, but from a bovine gene transferred into yeast cells. "Because the chymosin no longer came from suckling calf stomachs—and Britons are quite fond of their suckling calves—it was introduced as 'vegetarian cheddar' in 1991. The vegetarian organizations adopted it enthusiastically with their official stamp of approval," recalls John A. Bryant, a geneticist at the University of Exeter in the United Kingdom. One supermarket added a label to the product stating that it was produced using gene technology, then queried the first 1100 customers who bought it about their knowledge of said technology. Although 40% of the respondents claimed to know what "gene technology" meant, only 11% of them could explain it. Similarly, when GM tomato paste was introduced at about the same time in the United Kingdom—from tomatoes with a higher solids content that were therefore easier to process—the media and environmental groups tried to rouse public protest. The boycott lasted as long as it took consumers to realize that the GM tomato paste was 20% cheaper than nonmanipulated brands and tasted the same. People bought the GM product. When it comes to food, people also care about price.

Times have changed drastically, at least in the United Kingdom in the post–bovine spongiform encephalopathy era. Consumer pressure has rendered GM foods all but extinct, and the driving force appears to be fear. David Cove, a geneticist at the University of Leeds, had his students conduct a public opinion poll, with some astonishing results. They found a high positive correlation between level of education and willingness to try a GM food, but overall, more than 75% of the respondents said that they would not do so, citing fear of the unknown. "In the U.K., people have a fear of eating DNA! My wife went to buy tomato puree, and told the checkout girl that she was delighted to have found it. 'Why?' the checkout girl asked. 'Because it makes better lasagna,' my wife answered. 'Ah,' said the girl, 'it must be because they put DNA in it.' Well, we eat about 150,000 kilometers of DNA in an average meal! The public is certainly not very well informed," Cove says. As a joke, an organization called "Society for DNA Free Food," based in Australia, offers, for a high price, a recipe book of foods that lack DNA—they must not include organisms, as all food does. The organization's website (http://www.netspeed.com.au/ttguy) laments the lack of investigation into "the long-term effects of eating DNA." A survey published in *Seed Trade News* in December 1999 asked citizens of a dozen nations whether the statement "ordinary tomatoes do not contain genes while genetically modified ones do" was true or false, or if they did not know. Austria and Germany lost, with 44% of the polled populace believing the statement to be true. The United Kingdom had 22%, the United States 10%, and Canada 15%. The GM food debate is revealing a profound need for improved biology education!

A consumer backlash, even based on unfounded fears, can have a powerful economic impact, as researchers and farmers alike are now learning in several nations. The public interest groups that are fanning the flames of the boycotts on GM foods also directly attack the research that they claim is vital to testing for safety, adding violence to economic pressure. And this is where the Tulelake experience in the United States has resurfaced in the United Kingdom, at the Astra Zeneca Agricultural Research Center near London, and elsewhere. Sometimes the protests just don't make sense.

The victims of a brutal 1999 midsummer ambush by activists were two stands of poplar trees that had been genetically modified to produce less lignin, the cell wall component that must be removed in the manufacture of pulp and paper from wood. Lignin removal is costly and damaging to the environment. The European Union had funded the project, and in 1996, English Nature, the government body that advises on environmental matters, had determined that the experiment posed negligible risk to native species in the area. Another goal of the experiment was to better understand wood chemistry so that less chemical herbicide would be necessary to cultivate trees. Yet a group called the Genetic Engineering Network, purportedly bent on saving the environment, attacked the trees and then released a nonsensical statement accusing the company that had altered and planted them of displaying "contempt for our planet and the life it supports, including human life." But the protest, although destructive, did prompt the International Union of Forestry Research Organizations to publish guidelines to better educate the public about the nature of genetically modified trees. Low-lignin trees would save energy and reduce pollution, but the tabloids instead focused on the fear of genetic manipulation of trees growing close to a populated area. In other incidents, researchers have had their offices broken into and had experiments destroyed that had nothing to do with genetic modification, again indicating a lack of understanding of scientific matters among activists.

Anti-GM fervor jumped the Atlantic in 1999 when European grain importers began refusing to purchase GM crops—although both Americans and Europeans had been eating them since 1996 with no apparent ill effects. The environmental group Greenpeace kicked off the campaign by targeting the major baby food manufacturer in the United States, crafting a clever campaign to suggest that including genetically modified grains in baby cereal might somehow harm innocent infants. It wasn't long before Greenpeace targeted the makers of foods as diverse as cat dinners, chocolate, potato chips, and ketchup, then encouraged citizens to turn in their boxes of Kellogg's Corn Flakes in protest of being deceived into ingesting GM corn. Many manufacturers, aware that even the most beneficial and safe technology won't fly if the public perceives a risk and the market evaporates, began to talk out of both sides of their mouths, claiming on the one hand that GM crops are safe, but agreeing to remove them to satisfy consumer concerns—thus lending legitimacy to a conclusion that was never based on real scientific evidence. The fear that was spreading like an insidious infection was, as far as scientists knew at the time, groundless. This still holds true: No one has yet to fall ill from eating a GM food.

The language of a cover story in the magazine *Consumer Reports,* which eventually concludes and concedes that GM foods are probably safe, reveals a clear antitechnology prejudice. Next to a photo of a woman perusing supermarket offerings the article reads, "She doesn't know that the tortilla chips she just put in her shopping cart may have been made from corn whose genes were manipulated to kill insects." Many a reader who doesn't make it to the end of the long article would get the impression that people everywhere are growing second heads from eating GM foods. The article builds a case for labeling GM foods so that consumers can choose whether they purchase a GM food.

Current FDA policy does not require labeling products containing GM foods if they are determined to be much the same, in terms of chemical composition and how they are dismantled in the digestive tract, as nonmanipulated fare. That is, it isn't the source of a food but its constituents that are important in determining a possible health risk. For example, a natural variant of potatoes was found to contain a toxin, but a GM potato not. Under FDA policy, the natural variant containing the poison would be rejected, but the GM potato not, if it was indistinguishable chemically from a run-of-the-mill edible potato. However, by 2003, bulk shipments of grain may bear the label "containing genetically modified organisms" if 50 of the 130 nations that attended the Cartagena Protocol on Biosafety held in Montreal in January 2000 ratify the ruling by 2002.

The anti-GM food movement that began in Europe was catalyzed by efforts by large agricultural companies to force farmers into using certain of their products. The companies genetically engineered crops to resist their herbicides, so that farmers would have to buy the seed to use the herbicide. Some companies also, for a time, produced seeds with valuable characteristics but that were intentionally sterile, compelling buyers to purchase new seed each year.

But if any one event can be said to have truly electrified the GM food debate, it was a three-quarter's page report in the May 20, 1999, issue of the journal *Nature.* The peer-reviewed article had the attention-grabbing title "Transgenic Pollen Harms Monarch Larvae."

Although the Cornell University entomologist who wrote the paper, John Losey, called the study preliminary and many scientists quickly pointed out its limitations, protestors jumped on it, ignoring more than a decade's worth of published work demonstrating the safety of transgenic crops. Thanks to this brief paper, the monarch has become an icon, a mascot of sorts for the antibiotech movement. Many a protest these days includes at least one demonstrator in a butterfly suit, although the experiment examined larvae—caterpillars.

The oft-evoked monarch experiment reported in *Nature* concerns *bt* corn. *Bt* refers to *Bacillus thuringiensis,* a bacterium that produces a protein that destroys the stomach linings of certain caterpillars, such as the European corn borer, that can decimate a corn crop. For years farmers have sprayed the bacteria directly onto crops to keep the corn borers at bay, which required continued applications. But by stitching the *bt* gene into plants, the crops produce their own natural pesticide, albeit borrowed from bacteria. In 1998, 26 million acres of *bt* corn and millions of acres of similarly manipulated soy, cotton, alfalfa, canola, wheat, and sorghum were planted in the United States and found their way into a great variety of foods. However, the more constant presence of the toxin in GM crops raised questions about effects on the environment, which Losey's experiment was designed to test. The monarch was chosen to assume the role of the proverbial "canary in a coal mine," a familiar species whose decline would serve as an indicator of an environmental problem, although its population problems are well known to be mostly due to logging-induced habitat destruction in Mexico.

John Losey posed a simple question: What would happen if monarch butterfly larvae ate milkweed leaves that had been dusted with pollen from *bt* corn? The prediction was obvious—they wouldn't do well. The toxin is used precisely because it harms related caterpillars, it is in *bt* corn pollen, wind carries pollen, milkweed sometimes grows near cornfields, and monarch larvae eat milkweed leaves (Color Plate 5).

To model this scenario in the lab, Losey placed lab-raised caterpillars in plastic containers and fed them exclusively either milkweed leaves brushed with pollen from *bt* corn, leaves brushed with pollen from non-GM corn, or pollen-free leaves. The results were that 44% of the larvae consuming *bt* pollen died, compared with none in the control groups. The surviving larvae force-fed GM corn ate less and grew more slowly. In contrast, a similar study done at Iowa State University a year earlier showed very low mortality associated with eating *bt* toxin–coated leaves.

Many scientists pointed out that the simulation had little to do with reality. Said Anthony Shelton, another Cornell entomologist, "If I went to a movie and bought 100 pounds of salted popcorn, and then I ate the salted popcorn all at once, I'd probably die. Eating that much salted popcorn simply is not a real-world situation, but if I died it may be reported that salted popcorn was lethal."

Logic argued against the scenario too. Most milkweed plants do *not* grow near cornfields, and those that do tend *not* to attract monarch larvae. The butterflies are not endangered, apparently none the worse for the millions of acres of GM crops already cultivated. As one scientist pointed out, lawnmowers and pickup trucks cavorting through cornfields kill more butterflies than will ever be nauseated to death by *bt* corn pollen. Making the entire matter moot is that corn can be engineered *not* to secrete the toxin in pollen. Yet the Cornell study quickly became a classic, forever linking Losey's name with championing *Lepidopteran* rights. But it did raise a legitimate environmental concern, and research quickly intensified to retest Losey's hypothesis to confirm whether *bt* corn could realistically harm monarch or other larvae.

By early November 1999, some data were in, which researchers presented at a symposium held near Chicago. Various investigators had found that the concentration of corn pollen on milkweed plants growing in or near cornfields drops off very quickly with distance, confirming reports that the Environmental Protection Agency had consulted in 1995 when it approved *bt* corn. But the activists had ignored these earlier findings. Furthermore, corn pollen that does land on milkweed leaves is blown away or washed off quickly. Natural timing also did not favor a battle to the death between corn and caterpillars—95% of the Nebraska corn crop releases its pollen long before monarchs lay their eggs. And John Losey himself reported on experiments in the field demonstrating that butterflies do not alight on pollen-laden leaves! Caterpillars apparently have taste preferences. But because much of the research reported at the November meeting was funded by companies that manufacture GM seeds and pesticides and was presented to the press before it was published in peer-reviewed scientific journals, it drew suspicion. Environmentalists were so angry at the conflict of interest that they barely noticed the work reported. And so the image of a "toxic cloud of pollen saturating the Corn Belt," as one researcher put it, persisted.

Current fear of genetically modified foods seems to be spiralling out of control, having become more an economic and aesthetic issue than a scientific one. Soon after the Chicago meeting, an essay in *The New York Times Magazine* concluded that GM foods, although extremely promising, should be abandoned because of the scientific study that showed that they harm monarch butterflies! Like the children's game of telephone in which a message changes ever so slightly with each telling, Losey's original paper began to assume a life of its own, overriding anything that came after or had come before that argued otherwise.

The technical media proved just as guilty as the popular press in their hyperbole and misinterpretation of the one monarch study. In a late 1999 issue of *Science,* a "Policy Forum" article on the ethics of creating a simple form of life by a who's who of bioethicists concluded, "Nevertheless, novel species, however derived or introduced, are still cause for concern because they can wreak ecological havoc. Even small genetic alteration to organisms can have far-reaching, unintended consequences." It then cited the Losey paper. A few stunted lab-raised caterpillars, sickened by a substance already known to sicken them from decades of work, is hardly "far-reaching, unintended consequences." It seems the ethicists didn't actually read the paper, but subscribed to its widespread interpretation via a game of media-mediated telephone!

THE FUTURE

Only time will tell whether gene therapy and genetically modified foods will weather early warnings and setbacks to evolve into mainstay technologies. It is likely that they will. W. French Anderson deems gene therapy no less than the fourth revolution in medicine, following the realization that good sanitation practices can prevent infection; the use of anesthesia in surgery; and the advent of vaccines and antibiotic drugs to prevent and treat infectious disease. He predicts that by the year 2030, all diseases will be treatable at the gene level.

The coming genetic approach to medicine will consider several genes at once, as well as gene-environment interactions. With the influx of genome data for humans as well as our pathogens, this century will see the field of genetics broadened into genomics. Physicians will one day routinely consult genome profiles of potential patients to predict who will benefit from which specific treatment.

Out in the cornfield, genomics will accomplish what traditional agriculture has only sometimes succeeded in doing—identifying or creating gene combinations that improve crops. Consider one spectacular example of what mixing and matching genes from different species can accomplish: yellow rice (Color Plate 6), created by Peter Beyer at the University of Freiburg and Ingo Potrykus at the Swiss Federal Institute of Technology. This rice produces β-carotene (a precursor of vitamin A) and stores twice as much iron as unaltered rice. If the rice can be effectively delivered to farmers, it may go a long way toward preventing the widespread health problems of vitamin A and iron deficiencies. The new rice traits come from petunia, a fungus, green beans, and a bacterium, as well as boosted expression of one of its own genes. Yet this rice, like *bt* corn and low-lignin trees, still must be extensively tested in a natural habitat to learn its effects on ecosystems, and the gene combination must be transferred to readily cultivated strains. Delivery of this and other potentially life-saving GM crops remains a formidable hurdle to overcome, but one that extends beyond the realm of science.

Yet even the widely hailed yellow rice work has met the anti-GM controversy. *Nature* rejected the report without even getting it peer reviewed; it was instead published in *Science.* The fact that *Nature* comes from the United Kingdom and *Science* from the United States is hardly a coincidence, Potrykus has said publicly, implying that the British journal was bowing to political pressure. And the European Union denied funding for an additional year to the research team, which it would have used to transfer the genes to other varieties of rice to speed its utilization. Fortunately, other agencies are interested in funding the effort.

Many protests against genetically modified foods focus on combining genes from different species. But new gene combinations need not be evil. People have been using human insulin, growth hormone, and other drugs manufactured in bacteria since the early 1980s. Photographs of tobacco plants aglow with a firefly's bioluminescence protein have appeared in textbooks for years, more recently joined by mice that glow with a jellyfish's green fluorescent protein, used to label all sorts of organisms or their parts. Yellow poplar trees given a bacterial gene that encodes a mercury-degrading enzyme target mercury compounds in soil, releasing them in a less dangerous form from their leaves. Other trees, armed with a TNT-detecting bacterial gene and the glowing jellyfish gene, emit green light from their roots in a field riddled with land mines, indicating the locations of the buried munitions with a precision not possible using metal detectors.

The combinations and permutations of the genetic repertoire of life on Earth may be ultimately limited only by our imaginations. And this ability, this power, is perhaps why biotechnology frightens many people. Perhaps we are going too far, doing things that would not occur without our intervention. Other technologies that do not tinker directly with life, such as VCRs and computers, are not perceived with quite the same level of apprehension.

Is technology, such as gene therapy and genetically modified foods, the inevitable consequence of research into the way that life works? Certainly science does not typically start out with such a directed goal. Instead, the ideas that fuel a new technology emerge as basic understanding grows. Carl Woese's detection and description of a new variation on the theme of life answered and raised questions, leading to the discovery of microbes that thrive in extreme (to us) environments, suggesting that their chemicals might be of use (see Chapter 3). Leroy Stevens was interested in jumbles of abnormal tissue in mice—he hardly pictured the specter of warehouses of customized spare body parts that discussion of stem cells now evokes (see Chapter 6). And prions, the mysterious proteins that are normal parts of our bodies but that can somehow become trapped in a transient variant form that eats holes in our brains, may hold some practical use once we learn what they are and how they function and malfunction (see Chapter 4).

Scientific endeavor continually answers questions about nature and raises new ones, but it doesn't function in a vacuum. As science segues into technology, the factors of economics, history, politics, and sociology must be considered in the equation of consequences. And thus science ultimately affects every one of us. To see it as a series of disconnected breakthroughs or as a new tool or technology that surfaces from nowhere is dangerous. Only a scientifically literate society can develop technologies that do no harm or that balance risk with great benefit. And that literacy begins with understanding the continuity of discovery that is science.

REFERENCES

Anderson, W. French. "A cure that may cost us ourselves." *Newsweek,* 1 January 2000, 12.

Frank, Lone. "Consumer power heralds hard times for researchers." *Science* 287:790–791 (2000).

Hall, Stephen S. "One potato patch that is making genetic history." *Smithsonian,* September 1987.

Haslberger, A. G. "Genetic technologies: Monitoring and labeling for genetically modified products." *Science* 287:431–432 (2000).

Lewis, Ricki, and Barry A. Palevitz. "Science vs PR: GM crops face heat of debate." *Scientist,* 11 October 1999, 1.

Losey, John, et al. "Transgenic pollen harms monarch larvae." *Nature* 399:214 (1999).

Marshall, Eliot. "Gene therapy on trial." *Science* 288:951–957 (2000).

Niiler, Eric, "GM corn poses little threat to monarch." *Nature Biotechnology* 17:1154 (1999).

Palevitz, Barry A., and Ricki Lewis. "Fears or facts? A viewpoint on GM crops." *Scientist,* 11 October 1999, 10.

Salt, David E. "Arboreal alchemy." *Nature Biotechnology* 16:905 (1998).

Shelton, Anthony M., and Richard T. Rousch. "False reports and the ears of men." *Nature Biotechnology* 17:832 (1999).

Stolberg, Sheryl Gay. "The biotech death of Jesse Gelsinger." *New York Times Magazine,* 28 November 1999, 17.

Stolberg, Sheryl Gay. "Gene therapy ordered halted at university." *New York Times Magazine,* 22 January 2000, F1.

Stolberg, Sheryl Gay. "Agency failed to monitor patients in gene research." *New York Times Magazine,* 2 February 2000, F1.

Strauss, Steven, et al. "Forest biotechnology makes its position known." *Nature Biotechnology* 17:1145 (1999).

Ye, Xudong, et al. "Engineering the provitamin A (β-carotene) biosynthetic pathway into (carotenoid-free) rice endosperm." *Science* 287:303–305 (2000).

INDEX